ACCIDENT PREVENTION
AND LOSS CONTROL

ACCIDENT PREVENTION AND LOSS CONTROL

CHARLES L. GILMORE

AMERICAN MANAGEMENT ASSOCIATION

© American Management Association, Inc., 1970.
All rights reserved. Printed in the United States of America.

This book may not be reproduced in whole or in part
without the express permission of the Association.

International standard book number: 0-8144-5232-9
Library of Congress catalog card number: 70-116717

FIRST PRINTING

PREFACE

IT SEEMS strange that in our society an accident is always good copy for public consumption, while, at the same time, we are shocked at and defend against the injury or disaster that may befall us personally or our friends. In the complex structure of our industrial society, this human trait of concern has been transformed into active programs to prevent or minimize the effects of accidents to the employees and stockholders of businesses. Literally, millions of dollars are spent each year by American industry in these programs, which have many names, of accident or loss prevention.

Management is aware of the cost of these programs. The justification is only partially humanitarian. Managers are convinced that this expenditure reflects good, hard business sense. But, as with any other item of expenditure, managers want and deserve an evaluation of the effectiveness of this effort.

It is the purpose of this book to present a complete program for loss prevention based on methods that are in use and effective, so that management control mechanisms can be applied to improve performance. The first chapter is addressed to the manager who is responsible for the elimination or minimization of losses and must implement the loss prevention program. The second chapter is addressed to his safety specialist who is responsible for the measurement program that results in effective control. Success, of course, depends upon the joint effort of factual performance appraisal and firm, effective control.

Numerous individuals have contributed to the development of the concepts and techniques described in this book. To name a few is grossly unfair, but to name them all is impractical. The following were closely associated with major concepts: Mr. L. W. Buttery, formerly with Monsanto Company and now system analyst consultant for Bonner and Moore Associates, Inc., Houston, Texas, who with the author developed the basic personal injury measurement-control system; Mr. Tony Ozarchuk, loss prevention engineer, and Mr. John Y. Hutchison, engineering aide for Monsanto Company (Texas City) who set up the statistical program and made it function smoothly; Mr. R. L. Browning, loss prevention engineer for Monsanto Company (Texas City), who with the author developed the property loss prevention concepts; and, finally, the four successive plant managers at Monsanto Company (Texas City), J. S. Putnam, J. M. Chamberlin, W. R. Nisbet, and R. V. Butz, without whose faith and cooperation the entire concept would not have been tried. The contributions of these and of many others not specifically named are acknowledged with appreciation.

Charles L. Gilmore

CONTENTS

One	To the Manager	9
Two	To the Safety Engineer	12
Three	Existing Measurement Techniques	20
Four	The Anatomy of Accidental Loss	33
Five	The Performance Measuring System	39
Six	The Injury Report to Control	48
Seven	Extensions of Injury Effect Measurement	80
Eight	The Potential or Near-Miss Accident	114
Nine	The Property and Business Interruption Losses	120
Ten	The Loss Control Program	129
	The Author's Postscript	164
	Bibliography	165
	Index	171

CHAPTER ONE

TO THE MANAGER

THAT word "manager" implies a responsibility to control people and machinery to produce an end product or service. To meet that responsibility, a manager is interested in the prevention of losses to his people and facilities. Yet despite his busy schedule, he is willing to gamble that this book presents a new or different (but practical and proven) concept of loss prevention (profit protection).

Let us assume that you, the manager, have a real interest in loss prevention, and, frankly, it is hard to imagine a manager not being interested in performance and profits. However, in your organization, the "safety program" may be delegated to a committee or the safety department. Are the responsibilities for control of quantity, quality, and costs also delegated to a committee or a staff department? Perhaps you feel that safety performance is reflected only in personal injuries and as long as the "lost time frequency" is reasonably good, you would prefer not to be bothered. Perhaps some would define a "safety program" as some sort of social welfare program, something the organization cannot do without but not something to be taken seriously relative to the sophisticated science of important administrative activities. It is to be hoped that this would not be your definition.

How Important Is Loss Prevention?

There are numerous books on management methods and techniques, and all of them say, in one way or another, that your job is to control the productivity of people and facilities so that your company and your community profit from the existence of the enterprise. Accidental interruptions to the continuity of this business activity can only result in losses to your people, your company, and your community. Therefore, losses are failures of performance, and loss prevention becomes an important facet of your managerial responsibility. Simply stated, loss prevention is your business, and it is so important that you cannot afford to give it a low rank in your priority list of responsibilities.

The complexity of our changing industrial environment with its new, exotic products, competitive processes, and expensive equipment makes loss prevention more important. The changing management attitudes that accept the asset value of manpower and facilities and the increasing professionalism of the safety specialist give stature to this function. There is a revolution in progress among safety engineers who are no longer content to just fight fires, chase ambulances, and teach first aid. They are asking "Why?" and new tools are available to help management and the safety engineer perform this loss prevention function better.

You need evaluation and measurement techniques in order to control. Most industrial loss prevention programs have been oriented to the narrow concept of a personal injury prevention activity. In this area, the gains, according to the measurements that use fatality and disabling injury frequency and severity as a base, have been dramatic. But loss prevention includes the prevention of losses of facilities and production as well as personal injuries. Further gains in loss prevention are challenging, difficult, and require a scientific approach on a broader base. These loss problems demand definition to pinpoint the causes, to justify the effort, and to make that effort efficient. You want to measure progress so you know where you stand and where you are going. In loss prevention, effective management control has been waiting for an effective and sensitive method of measurement. This is exactly what this book is about.

Control Through Measurement

What is the concept of this loss prevention management control system? Losses are cause-and-effect situations. Effects are the result of—the reason for—the definition of the loss sequence. Since there is an effect, performance can be measured. Since there is a cause, performance can be controlled. Control depends on the ability to eliminate or minimize the causes of loss.

This book develops a complete program for loss measurement based on actual and potential effects and offers circumstantial analysis as a replacement for loss occurrence, thus laying the groundwork for more effective management control. This program employs the existing accepted measurement and statistical techniques of other disciplines and adapts these for loss prevention performance measurement. This adaptation takes the guesswork out of loss prevention, reorients it away from crisis attitudes, and places it on an analytical scientific level with other management control systems.

Management systems are effectively used to control quantity, quality, and costs. Now there exists a similar device for the control and reduction of loss. The description and techniques of this measurement method as a tool for better management control are presented to you and your safety engineer for evaluation and adoption in your business.

CHAPTER TWO

TO THE SAFETY ENGINEER

OVER the years, you have worn many hats. The position of safety engineer was created out of the humanitarian and social pressures resulting from personal injuries, the economic pressures arising out of fires and explosions, and the legislative pressures of regulatory inspection. Safety engineers have fostered safety programs, headed campaigns to motivate safety consciousness, directed the fire protection effort, guarded the plant security, taught first aid, sat with the injured, administered workmen's compensation insurance and other insurance programs, participated in community, state, and national safety meetings and organizations, issued personal protective equipment, lectured to the local PTA and pleaded for school safety, and if they were not already worn out, participated in civil defense, mutual aid, and civic projects.

Past Performance

You have used every device you know or have heard of in your dedicated crusade to make the job safer. These approaches have become traditional to your business: the educational effort of training, publicity, and promotion; the rewarding of effort through plaques, prizes, and stamps; the post facto investigation of incidents in the hope of preventing recurrence; the meeting of standards set by someone else, often at a level that did not satisfy you; the working with people on committees, in which you did most of the work. Through it all, you probably sounded like John the Baptist as you withstood the pressures of production and the scarcity of money.

And the reductions in disabling injury frequency and severity are a compliment to your effort. This has been your prime area of effort in loss prevention because most safety programs have been limited to the prevention of personal injuries and the provision of fire protection. Safety engineers haven't been involved when a machine broke down unless someone was hurt. You weren't consulted about a production curtailment unless it was the result of a fire or an explosion. You were only called upon if an injured man had to be cared for or a fire had to be put out. But loss prevention involves considerably more than this.

Traditionally, when a loss (human or property) has occurred, safety engineers have acted to give care (medical treatment or fire protection to limit the loss) and to investigate the loss. This investigation resulted in action to prevent recurrence of this loss and generated a description of the loss, its cause, and corrective action. This approach has been used for years and has accomplished some loss prevention, but it has many inadequacies and shortcomings.

First, it has been primarily limited to personal injury control, and personal injuries are not the only loss resulting from accidents.

Second, we have allowed the loss severity rather than the loss capability to predicate the amount of attention given to the accident investigation. Prime examples of this are the cursory examination of accidents leading to minor injuries and the unlimited activity aroused by an injury that results in lost time. There is no real analytical examination of the capability of the accident.

Third, we have further compounded the problems of loss prevention by substituting protection for prevention. We have not recognized probability but accepted possibility as the only criterion for action. To engineers, this is an unworkable thesis.

Fourth, these investigations have tended to ignore the loss

incident because tunnel vision has concentrated the effort on the particular loss rather than the capability of loss.

Loss prevention has to do with the prevention of accidents, all accidents, not just those that result in injury or damage. This is a broader concept and the one that managers are asking safety engineers to recognize.

Management, in its other functions, has developed management control systems for quantity, quality, and costs and recognized the need for a similar system to control losses. Although the protective activity associated with safety is practiced, a more effective preventive effort is needed. And since the safety of people and facilities are involved, management is looking to the safety engineer. It is time for a more positive approach. For example, those measurement methods that enable the safety engineer to understand and predict what can occur and plan controls before losses occur must be developed.

A Growing Professionalism

There is a growing professionalism among safety engineers. (This does not imply that you weren't professional in the past but simply that you know more today than you did yesterday, and you are going to have to know much more tomorrow just to cope with the problems and work in the environment.) This is apparent in the following quote from a brochure prepared and distributed by the American Society of Safety Engineers:

> The safety professional brings together those elements of the various disciplines necessary to identify and evaluate the magnitude of the safety problem. He collects and analyzes the information essential to the solution of the problem. He is concerned with all facets of the problem, personal and environmental, transient and permanent, to determine the causes of accidents or the existence of loss producing conditions, practices or materials.
>
> The safety professional in performing these functions will draw upon specialized knowledge in both the physical and social sciences. He will apply the principles of measurement and analysis to evaluate safety performance. He will be required to have fun-

damental knowledge of statistics, mathematics, physics, chemistry, as well as the fundamentals of the engineering disciplines.

He will utilize knowledge in the fields of behavior, motivation, and communications. Knowledge of management principles as well as the theory of business and government organization will also be required. His specialized knowledge must include a thorough understanding of the causative factors contributing to accident occurrence as well as methods and procedures designed to control such events.[1]

This growth of the safety engineer is no indictment of the past, only a challenge for the future. The professional approach merely applies a mixture of logic and new techniques to the science of problem solving. The traditional approach has centered its effort on current accidents, set up guards to minimize the result, and developed programs to protect against the effect of the accident. The scientific approach, long applied to the research and engineering of quantity, quality, and costs, centers its effort on why the accident occurred, with only secondary interest in the ultimate result. In contrast to the traditional, *protective* approach, the scientific approach is *preventive* in nature, seeking the solutions to problems before they become losses. The protective method assumes that the accident must inevitably happen although the loss can be limited. The preventive technique assumes that the causal sequence can be recognized and aborted, thus saving lives and facilities. Accident prevention has no foundation if it lacks scientific and engineering fundamentals.

Space-Age Techniques

Perhaps the one single event that has created this emphasis on the scientific has been the advent of the missile and space age. Scientists who worked with missiles and spacecraft could not afford even the first failure. Industry can use the concepts and techniques of this age to improve its own performance. Researchers in accident

[1] "Scope and Functions of the Professional Safety Position," published by American Society of Safety Engineers.

prevention are now seeking answers to questions about why men and machines behave as they do. We seek to define the *reliability* of man and his machines. As long as men and machines must work together to produce the things and services we demand, there will be accidents to both the man and the machine. But by working to prevent those occurrences, we are improving the reliability of both men and machines and reducing the probability and severity of the accident. Research gives us an understanding of why—a necessary ingredient of prevention.

Our tools are improving, both in quantity and quality. Personal safety equipment has become more useful and effective and, less important, more stylish. Fire equipment and methods of first aid are improving. The ability to record and analyze data has become a science in its own right, and the hardware and methods are becoming more versatile almost daily. Analogs are used for solving problems such as the hydraulics of fire-water systems, cutting engineering time from months to minutes. Other tools, such as control charting and logic analysis, can be used to prevent accidents just as they are presently being employed in various other disciplines.

Improved technology has changed quality control from a process of obtaining a product sample eight hours after production—a mix of a little of this and a pinch of that—to a precise science of statistical sampling and mixing. Management exercises control of quality through measurement parameters provided by staff quality specialists. Cost control is now exercised through computerized cost sheets as interpreted by cost control specialists. Management is experiencing a better decision batting average with these measurement-control techniques, and it is showing a profit on the cost sheet. The manager has gained confidence that decisions for control based on scientific measurements will be right more times than those based on the old guess-and-gamble techniques. He wants the same reliability in the measurement of safety performance, and he looks to the safety engineer for it. He isn't asking any more of him than he asks of other functions. He wants improved performance through an efficient effort expended in a justified program. He wants and deserves a definitive appraisal of the seriousness of the problem, and he wants justification for the effort safety engineers want him to make.

The Way to Improved Performance

This desire is shared by the safety engineer. With all the various tasks that already crowd his schedule, he is also interested in defining the problems and assessing the severity so as to pinpoint where the emphasis is needed and where the effort will be most valuable. A measurement of loss prevention performance is needed as a basis for making recommendations regarding improvements. With this measurement, safety engineers can capture the attention of the manager. They can inform him of the status of the loss prevention effort, not by pointing to a lot of possible problem areas in the hope that the shotgun blast will somehow hit the real problem, but rather by defining specifically the soft spots in the effort. Having done this, the need and justification for the effort required is apparent. The safety engineer can confidently expect that the effort will succeed because the program is not diluted by dead-end ideas and wasted time. The capabilities of those responsible for execution of the program are concentrated on a target, so the program is efficient as well as justified. The safety engineer will gain a professionalism in the eyes of the manager and raise the stature of both his program and his function.

Perhaps the safety engineer believes he is already providing a good measurement of performance to the manager. He tells him about the disabling injury frequency and severity. He warns the manager that this or that is a problem and that something ought to be done about it. Anyway he is busy with investigations, fire protection, housekeeping, and a multitude of other duties. He doesn't really have time to do all this "measurement" stuff and really doesn't understand all those fancy computer programs. But perhaps he would do well to stop running after the crisis of the day and look at his job objectively. He may find that many of the things he does now really do not need to be done at all.

Take investigations as an example. Why participate in them? For one's own information? To show the importance of the activity by one's presence? To referee between the accuser and the accused? The truth might be that the safety engineer is in the way, that he confuses the issue, that it's really none of his business. The loss, probably an injury, resulted from an accident that involved at least the injured person and possibly others. If the safety engineer is not responsible or accountable for either the department or the people, why is the engineer involved in the investigation? Can he perform the investigation better than the supervisor who knows the department and his people? Isn't the supervisor concerned enough to seek the truth? If he isn't, the plant has another problem, and it won't be solved by the safety engineer's presence at the investigation. When the engineer becomes the crutch on which the supervisor leans, one of the supervisor's greatest opportunities to *supervise* is taken away from him. He is denied the opportunity to seek the cause of the accident and correct that cause and is thus denied the ability to raise his stature among his people because he has shown concern for his people and facilities. Let him supervise, let him be the leader, let him stand for safety in front of his people. Every safety engineer should ask himself: Am I doing my job or someone else's?

The first responsibility of a safety engineer is to evaluate the effectiveness of management's program for loss prevention, primarily because he is the only man in the organization who can make an objective evaluation. All other functions are prejudiced by conflicts of interests—managers want all functional performances to be good, production people have production priorities and pressures that tend to deemphasize safety, and so on. Management's program is to be evaluated—not the safety department's program. We are talking about a management responsibility—loss prevention—and it is this performance that is to be evaluated. The safety engineer's performance might well be evaluated according to how effectively he measures this loss prevention effort.

This measurement or evaluation may be made in many ways. A record of the number of safety meetings held may be kept and, by attending a few of them, the safety engineer can determine their quality. He may visit various departments and make spot inspections just to see how things are going. He may receive investigation reports in order to evaluate how thorough the investigation has been, how appropriate the "action to be taken" will be, and whether and how quickly the action was taken. But this is all done as measurement of performance, not as a part of someone else's job. The safety engineer develops the measurements and then interprets them to ascertain the current status of performance and to define problem areas. He thus builds confidence in the measuring device and in his own interpretations. Then he recommends a program that is a reasonable and justified solution to the soft areas of performance. The program itself will be management's responsibility, because success comes only from management's acceptance and authority. But the safety engineer can now return to measurement—to monitor the progress of this new effort.

His role should be that of a "systems evaluator, specialized in identifying errors in the management system through loss incidents . . . not . . . a social welfare worker operating in an organizational vacuum." [2] No other responsibility can profit the company more than the safety engineer's ability to measure loss prevention performance so that management can make the timely, efficient, and effective decisions necessary for the control.

This book presents a complete measurement program, a tool to enable management to direct effective control of real or potential loss at causes rather than effects. The concept of this program is that accidents are cause-and-effect situations, that the effects, both real and potential, provide the base for measurement of performance, and causes, once recognized through the measurement technique, provide the direction for control.

[2] W. C. Pape and T. J. Creswell, "Safety Aids Decision Making," unpublished report.

CHAPTER THREE

EXISTING MEASUREMENT TECHNIQUES

SINCE the advent of the safety engineer, measurement on some basis became necessary. It is interesting that the one measurement device with any widespread acceptance—Z16.1 Code drawn by the American Standards Association—was intended only to record the injury experience of industry in a standard method, not to measure performance. It was written to provide a standard definition of a reportable work injury and a standard schedule for the severity of the injuries to be recorded under this code. It also provided the mathematical method for calculating disabling injury frequency and severity. This widely accepted code has been around for many years, and revisions have been generally limited to clarifications or expansions of intent and interpretation. But the intent was to record injury experience, not to measure performance.

Z16.1 Code

As a thirsty man offered a drink of water, the safety engineer and subsequently the manager eagerly latched onto the code. Here at last was a measuring device for comparing safety performance. As this measurement of performance, which used fatalities and disabling injuries for the base data, was accepted by most industries, it became the basis for competition between plants within a company and among companies within an industry. The National Safety Council provided coveted plaques complimenting plants and companies for a disabling injury frequency rate of zero over periods of millions of exposure hours or a calendar year. Awards of varying values were given to plant employees to acknowledge these periods of safe operation. However, one company's statisticians calculated that the odds were 12 to 1 that the length of time between disabling injuries would not be doubled, indicating the difficulty of attaining and maintaining safety records.[1]

The National Safety Council has compiled a table summarizing the combined experience of all industrial reporters to the Council from 1926 to the present.[2] The results are impressive. First of all, the number of units reporting has increased from 1,725 in 1926 to more than 10,000 in 1968, thus indicating the interest of industry in this business of safety. In the Council's tabulation, three statistics are recorded. In the 40 years recorded, the disabling injury frequency (number of disabling injuries per million man-hours worked) has decreased from about 30 to 7. Similarly, the disabling injury severity rate (days lost per million man-hours worked) has been reduced from about 2,500 to less than 700. The third statistic recorded, the average time charge (days lost per disabling injury reported), has not substantially changed over the 40 years, remaining at 100 days lost per disabling injury. This statistic could be called the "severity factor." The consistency of this severity factor indicates a consistency of reporting among all industry in addition to a steady reduction in both frequency and severity rates. This is a remarkable record.

It is interesting to note that the records of the safety leaders in industry show low frequencies but sizable severities, thus yielding a severity factor that is considerably higher than the National Safety Council's reported industrial average. This would lead to the conclusion that only the more severe disabling injuries were accepted as reportable under the Z16.1 Code. Perhaps the interpretation of the code has been influenced more by competitive reasons

[1] R. K. Mueller, *Effective Management through Probability Controls* (New York: Funk & Wagnalls Company, 1950), p. 177.
[2] National Safety Council, *Work Injury Rates*, 1969 Edition, p. 26.

than by the standard of sound performance measurement. The safety performance of these corporations is not poor; quite the contrary, they are undoubtedly leaders in accident prevention. But as the Z16.1 Code became accepted as the only standard for both measurement and competition, corporate managements in their necessary function of control demanded improved performance. Local plant managements recognized this pressure to effect a reduction of the disabling injury frequency. As a natural result when the expected safety performance was known, the pencil was sharpened, and the code was interpreted with an eye on this performance goal. Naturally, a "winner" becomes the target for charges of misinterpretations of the code and unfair practices such as allowing the injured to answer the plant first-aid station telephones or bringing in stretcher cases from their homes to avoid a "lost timer." Whether any of these charges are true is doubtful, but the possible misuse of the measurement device has taken away some of the value of the Z16.1 Code.

Russell DeReamer [3] felt that disabling injury rates are indicators of progress in accident prevention only when the man-hour base is sufficiently large. While these rates may be a fairly good indicator for plants or companies, they have little value as a measure of the safety performance of a supervisor. Many present-day safety engineers recognize the limitations of using the disabling injury as a basis for measurement of performance, and some corporations no longer use it as the basis for their measurement of performance. William E. Tarrants has stated that "present attempts to control accidents and their consequences can best be described as trial and error chiefly because adequate measures of the effectiveness of this control do not exist." [4] Thomas H. Rockwell believed that "because accident frequency and severity are so susceptible to chance, one can have little faith in accident frequency and severity rates as true measures of system performance." [5]

John V. Grimaldi felt that safety performance was indicated in severity rates and average time charged rather than the generally

[3] Russell DeReamer, *Modern Safety Practices* (New York: John Wiley & Sons, Inc., 1958), p. 294.

[4] William E. Tarrants, "Applying Measurement Concepts to the Appraisal of Safety Performance," *ASSE Journal,* May 1965, Vol. X, No. 5, p. 15.

[5] Thomas H. Rockwell, "A Systems Approach to Maximizing Safety Effectiveness," *ASSE Journal,* December 1961, Vol. VI, No. 6, p. 21.

accepted frequency rate. He states that "frequency rate, as a measure, more properly reflects injury control, not accident control." [6] A. C. Blackman reports that "a recent study of our own membership (American Society of Safety Engineers) developed the point that there was no relationship between frequency of injury and severity, as computed by the ASA Z16 standard." [7]

But safety engineers were and are searching for the "adequate measure," and they have developed different measurement techniques using the disabling injury as a base.

The Disabling Injury Index

A. C. Blackman described a Disabling Injury Index as the mathematical product of frequency and severity rates divided by 1,000 and asks if this is a valid and reliable measure. There are those who must feel that this index is (or may be) valid and reliable because it is included in the newest revision of Z16.1 as a nonstandard measure. This was the result of requests for a combined frequency–severity measure.

Western Electric has devised a single safety index.[8] The formula is as follows:

$$SSI = \frac{C \times 1,000,000}{16 \times D \times P}$$

where P = Number of plant personnel.
D = Days in the period being measured.
C = Charges as the total number of calendar days lost in excess of the first seven calendar days or 10 percent of Z16.1 schedule (whichever is greater) for both on- and off-the-job injuries.

Four new concepts have been added to the disabling injury frequency method in this formula. First, although it uses the definition of the disabling injury, the index depends on the severity of the

---
[6] John V. Grimaldi, "Another Look at Stimulating Safety Effectiveness," *ASSE Journal*, April 1962, Vol. VII, No. 4, p. 20.
[7] A. C. Blackman, "Professional Safety Engineering—A Need and Opportunity," *ASSE Journal*, May 1962, Vol. VII, No. 5, p. 26.
[8] E. J. Schowalter, M.D., "A Year's Trial with a New Safety Measurement Plan," paper presented at the Ninth Annual Western Industrial Health Conference, Oct. 9, 1965.

accepted injury. Second, the off-the-job disabling injury is given the same weight as the on-the-job injury. This recognizes the effect of the nonoccupational injury that keeps the worker from his job. Third, the count on days charged begins with the eighth day of disability. This is equivalent to the waiting period in many compensation laws. It is also a change that has been proposed in the Z16.1 code by some medical personnel as a means of removing the doctor who treats the injured from the pressures of interpretation. Fourth, the use of 10 percent of the severity schedule for days lost under the Z16.1 code recognizes what some consider inequities within that schedule.

Although some may feel that such a formula is playing with numbers or aborting the intent of the code, the concepts are interesting. A test of a plant's performance by this index might serve to make managers more aware of the costs of manpower losses due to accidents. It doesn't really make much difference where the accident occurred, since the productivity of the worker is lost to the company until he is able to return to work. The index seeks to give a broader base to the accident prevention program; it includes only those injuries from which the worker loses more than a week from work, the type that is usually included under the Z16.1 code. And these injuries are generally covered by either group or workmen's compensation insurance, indicating that the injury is considered to be deserving benefits under a different set of rules and schedules.

The Safe-t-Scores Method

Another approach that uses injuries as the base for measurement is the Safe-t-Scores method. James A. Martin and Dr. Gordon B. Wheeler of Pratt and Whitney Connecticut Atomic Nuclear Engine Laboratory developed the Safe-t-Scores because they felt the Z16.1 frequency and severity rates "are not reliable measurements of safety performance. . . . The measurement of safety performance should eliminate as much of the luck factor as possible, and should completely disregard any events or actions that take place after accidents." [9] The base of their measurement method is the all-

---

[9] J. A. Martin and G. B. Wheeler, M.D., "Safe-t-Scores: A New Measurement for Safety," *Occupational Hazards,* April 1962, Vol. 24, No. 4, pp. 35–39.

occupational accident frequency rate. Martin and Wheeler define an accident as "a preventable [occurrence] which in the opinion of the safety engineer could have resulted in a disabling injury." This means that the numbers used depend on the judgment of the safety engineer to objectively classify accidents so that the measurement is not affected by opinion. "This demands experience, careful investigation, knowledge of plant jobs and working conditions, and a lot of common sense." [10] The formula is as follows:

$$t = \frac{F_c - F_b}{\sqrt{F_b/M_c}}$$

where t = Safe-t-Score, a dimensionless number indicating poorer performance when positive and better performance when negative.
F_c = All accident-injury frequency rate for current period.
F_b = All accident-injury frequency rate for a base period, recommended to be a settled period of 12 months.
M_c = Number of man-hours worked in the current period in million man-hours. The frequency rates are expressed as the number of accidents (count an accident only once even though it involves several injuries) per million man-hours.

The developers of this formula claim statistical significance—that its results are an effective tool for management. They feel that it is the evaluation of the difference between the (current) rates and the standard which is important. They also state that this measure instantly indicates significant changes in performance. A score beyond +2 automatically shows that performance is poor enough to require corrective action, and, conversely, a score beyond −2 indicates a significantly better performance.

Some groups have designated F as the total injury frequency rate without regard to potential. In so doing, they feel that a more objective picture of the actual performance is obtained because interpretation is minimized.

Robert Brenner and J. H. Mathewson, feeling that severity rate had "no semblance of meaning to management from the standpoint of production," developed the concept of "cumulative effect of accidents, or CEA." [11] Time away from the job is production data

[10] Ibid.
[11] Robert Brenner and J. H. Mathewson, "The Principle of Accident Effect Reporting," *ASSE Journal*, January 1963, Vol. VIII, No. 1, pp. 9–14.

that should be meaningful to production decision making. Therefore, a daily record of the cumulative absence from work by injured employees would show periods of high absence due to occupational injury and would affect the planning of production manpower needs as well as indicate periods of poor safety performance. At the time of their paper, this was an untested concept, but it did show another attempt to use disability numbers as a tool for management control.

In the search for a measurement-of-performance device, these and other studies have utilized the disabling injury frequency and severity rates. But the need for a large exposure base in man-hours and the relative infrequency of the incidents have limited the meaningful use of such studies to industries or large corporations. The controversy over the value of these data arising from questions of interpretative integrity and competitive pressures seems academic when it is realized that in most plant subdivisions, disabling injuries are so infrequent that the data become statistically insignificant anyway. This does not mean that the injury is any less significant or that any plant's injury record is any less creditable. It simply means that there is a need to develop a better performance measuring system. This improved system must recognize that the accident prevention program includes accidents, not only those that result in personal injuries but also those that result or have the potential to result in any loss. This system must measure current performance with significant data and provide indicators so that management can promptly recognize change and initiate necessary effective corrective action.

The Serious Injury Classification

A most significant step forward was made in injury-base measurement techniques in the mid-1950s when Klingel and Haier of the Standard Oil Company of Ohio set up a serious injury classification which provided a more significant data base for measurement. According to their index, serious injuries include those resulting in temporary total disability, permanent disability, transfer to duty other than injured's regular work, fractures, lacerations requiring sutures, and eye cases requiring treatment by a physician. This classification and the measurement involved have certain real advantages over the disabling injury frequency methods. It provides more data, thereby increasing the significance of the relative numbers obtained. The safety engineer is no longer the interpreter of which injuries are reportable. X-rays and the medical judgment of the doctor, motivated only by the necessary and proper care of the injured, are the only tests. Emphasis is placed on the injury, whether or not the injured appreciates the particular recording system and whether or not the supervisor is sensitive to the injury reporting, simply because the injured is hurt seriously enough to demand medical attention.

The Serious Injury Index, which is in fact the frequency rate of the serious injury, is a simple, easily understood standard not particularly subject to pressured interpretations. If the safety engineer can maintain the integrity of this index and withstand the great temptation to play games with this device, the Serious Injury Index can be an extremely valuable tool for the safety engineer and manager. Olin Mathieson plants use a Serious Injury Index to gauge the effectiveness of safety programs when the traditional disabling injury base measurements seem to stop registering. Other companies are also using this injury classification as a base for measurement activity, and the most recent revision of Z16.1 describes this classification as a "nonstandard measure." This index will be discussed in considerable detail, and statistical applications that use injury data will be shown in the chapter on loss statistics.

Sampling

Another area of measurement that has been explored by the safety engineer is the technique of sampling. Work sampling has been successfully used for years by the industrial engineer. It has proved to be a statistically acceptable method of examining a small fraction of an activity to give an accurate picture of the whole activity. Since sampling measures the current level of activity, it, in effect, indicates what the activity or practices will be unless a change is made. For instance, if work sampling shows that workers are taking 30 minutes for the smoking break when the break is supposed to be limited to 10 minutes, it can be assumed that they will continue to do so unless the work practice is changed. This is true also with safety practices. If a worker uses Channel Locks instead of the proper wrench for a particular job, he will continue to do so until the practice is changed or an accident occurs. Sampling can be used to recognize a poor practice and motivate correction before the accident occurs.

Safety sampling has been used in several ways. One of the better known and undoubtedly more successful safety sampling programs is the Du Pont Safety Control Program. In this program, the number of unsafe acts and unsafe conditions (that could result in injury) found in a random selected sample of a work area indicates a specific measure of the safety level in that area. This sample was made in each work area on a once-per-week schedule by a team of supervisors. The inspection was limited to 15 minutes. The inspection time and members of the inspection team were selected at random. The results of the inspection were plotted on control charts for easy interpretation. Since both the method of sampling and the method of charting the results are well-known and accepted statistical procedures, the program would give factual evaluation of the area performance. In order to insure both uniformity and competence of the samplers, training was provided to all supervisors in the program. Du Pont feels that the program has proved successful in its primary purpose of appraising the safety level of individual areas and groups of employees within their plants. Because of the random timing of the inspection, the plant remained constantly alert, and the general safety level was improved. Supervision was better trained to observe safety infractions and conditions, and correction was more timely with improved safety control. The employee was interested and could see the interest of management demonstrated. It aided in the promotion of safety in both supervisors and workers.

In recent years several companies have used similar work sampling techniques. Thiokol Chemical Corporation at the Longhorn Ordnance Works, Chrysler Corporation at the Stamping Plant in Ohio, and Monsanto Company at several of its plant locations, to name but a few, have used sampling successfully to appraise the safety level of groups within a plant. In addition, teaching supervisors to observe the safety practices within their authority has proved to be a real asset to the general upgrading of safety awareness and improvement of work safety.

At the Tennessee Operations of the Aluminum Company of America, a safety observation program was developed along methods similar to the Du Pont program. It included an eight-hour supervisory training course followed by field sampling for unsafe acts. These sampling data, together with injury data and data obtained by work sampling for efficiency and productivity, were exhaustively studied. It was concluded that, although the program was not an effective factor in reduction of injuries, it did highlight safety problems and reduce unsafe acts—a process that might ultimately lead to a reduction of injuries.

Housekeeping inspections and managerial participation through field inspections and contacts are also methods of sampling. Although they are not usually statistical in nature, these methods have tended to upgrade the performance of a plant or operating unit.

Another form of sampling has been used on accidental occurrences. Not all accidents result in injuries, nor do all accidents have the potential to result in a disabling injury or serious loss. William W. Allison has proposed High Potential Accident Analysis as a concept of accident prevention. He defines high potential accidents as "those that did or under similar or slightly different circumstances could result in serious injury or damage." [12] If the accident had been recognized as having a high potential of causing loss and correction had been made, it should have been prevented. This approach requires the investigation of incidents and near-misses so that those with high potential can be separated from those with low potential. Having done this, the safety engineer and manager can concentrate on the correction of these high-potential hazards. This tends to improve the efficiency of the safety effort, since emphasis is placed on the more serious consequences of the accident.

William E. Tarrants, in his development of the Critical Incident Technique, has used a sample group interview to identify unsafe acts and conditions which contribute to accidents. The interviewer questions a group of people who perform particular jobs within certain work environments and asks them to describe unsafe acts

---

[12] William W. Allison, "High Potential Accident Analysis," *ASSE Journal,* July 1965, Vol. X, No. 7, p. 10.

and conditions they have observed. It is immaterial whether the incident actually resulted in injury; the only important point is whether such a consequence is possible. Tarrants has concluded that since causal factors leading to accidents are identified and this method reveals more information about causes than other available methods of accident study, a more sensitive measure of total accident performance is obtained. Tarrants also concludes that since this technique "is sensitive to accident problems which have a potential for accident loss but have not yet produced a loss—we are now able to identify and examine our accident problems 'before-the-fact' instead of 'after-the-fact' in terms of their injury producing or property damaging consequences."[13] In the study using the Critical Incident Technique, 52.1 percent more unsafe acts and conditions were identified by this technique than were identified by the accident records. Also, almost 75 percent of the different incidents reported were estimated to have occurred with a very high repetition frequency, even as often as every day, indicating a tremendous exposure to potential injury-producing accidents.

Human behavior has been studied with sampling techniques by universities and industry. It has been found that workers behave unsafely a significant proportion of the work time. One study conducted in industry showed this level to be about 20 percent of the time. When workers were closely supervised, this percentage dropped to about 6 percent, only to return to the normal level when close supervision was removed. With the exposure indicated by this level of unsafe behavior, it seems indeed fortuitous that more injuries of major consequence do not occur.

One program aimed at improving human behavior that has been widely promoted and applied in industry in recent years is the Zero Defects Program. Here the individual is encouraged, through publicized promotional reminders and training and supervision, to avoid errors in quality, production, and safety. Although the goal of zero defects is improbable, the workers are given a high standard of performance to attain. But the goal is similar to the goal of zero disabling injury frequency and severity. Therefore, the Zero Defects Programs are plans for motivating improved performance, not for measuring performance.

[13] William E. Tarrants, "Applying Measurement Concepts to the Appraisal of Safety Performance," *ASSE Journal*, May 1965, Vol. X, No. 5, p. 21.

Systems Analysis

A different form of performance measurement has resulted from the space age and has greater application to machines than to people. Systems analysis uses the probability of a failure or a series of failures to predict the probability of the unwanted consequence. By detailed analysis, it identifies the most probable path of failure and focuses attention on the corrective actions needed to reduce or eliminate the probabilities of this path. But systems analysis has not been widely used by safety engineers for the following reasons:

1. Prediction of mechanical or electrical device failure is fairly accurate, but prediction of human behavior is more difficult.
2. Safety (involving humans in relation to equipment) is not a go–no-go situation of failure or success.
3. Traditional safety personnel are reluctant to accept the concept of degrees of safety. The idea of risk-profit tradeoff, which is implicit in every management decision where risks and profits are in contention, is incongruous with the zero disabling injury frequency rate and other traditional systems.

Where systems analysis is used, the measurement of performance is the probability of the occurrence of the unwanted event. Having explored the various paths of failure and obtained a probability of failure together with an estimate of the loss damage, managers can decide whether the effort to prevent the failure is justified. The penalty of failure must be evaluated against the cost of preventing or minimizing the failure. Systems analysis is a highly sophisticated form of performance measurement. It apprises management of a consequence—its probability and its severity—thus making a logical decision possible.

Two methods of systems analysis can be briefly mentioned. The Fault Tree Analysis was first conceived by Bell Telephone Laboratories in 1961 and was used extensively by the Boeing Company and its associate contractors on the Minuteman Program. It uses a logic diagram, beginning with the undesired event, and develops the various paths of sequences of occurrences which could lead to that event while searching for the more probable paths. The Failure

Mode and Effect Method uses a similar logic diagram but begins with a component failure and works through the consequences of that failure to predict the ultimate probable undesired event. There are, of course, advantages in certain situations of one approach over the other, but each has its application.

George Peters and Frank Hall have been attempting to establish a method for analysis and control of human error. They frankly admit that inadequate data of the probabilities of human error in performing various work activity is limiting progress in this field, but as more data are accumulated, accident prevention can be measured and controlled in a new dimension. Certainly if human work behavior probability can be so quantified, the engineering and education effort can be directed to reduce the probability, frequency, and severity of the undesired injury.

In summary, safety performance measurement has been receiving a lot of attention. Basically, there are three areas in which this work is being carried out: the frequency and severity of the recorded accident, the sampling of human behavior in regard to unsafe acts and conditions, and the determination through logic and reliability data of the probability of the undesired event. This effort can gain a professionalism for the safety engineer, because it is his function to measure the performance so that the manager can control and improve that performance. There has been no attempt to give either a complete or an exhaustive analysis of any particular method of measurement. The described techniques have been presented only to apprise the safety engineer of the activity in this important function.

CHAPTER FOUR

THE ANATOMY OF ACCIDENTAL LOSS

BEFORE safety engineers can measure the effectiveness of the loss prevention effort, they should understand the process by which an accident occurs. This loss process is a series of sequential causes and effects which, if not arrested, ultimately result in a loss. The loss process is a cause-and-effect situation, consisting of three distinct phases—the loss exposure, the loss incident, and the actual loss.

The Loss Exposure

The loss process begins with an exposure—the prerequisite of the undesired loss. The loss exposure is a static condition or group of conditions that have the potential, under certain unplanned circumstances, to proceed to a loss.

To be significant, the loss exposure must include both the capability to cause and the capability to sustain a loss. This capability to cause the loss is the capability to damage or destroy property, to injure or kill people, and/or to interrupt or restrain a normally controlled activity involving people and/or machines. The capability to sustain the loss simply requires that there be something to damage or to interrupt. It is the value of persons, property, or activity subject to injury, damage, or interruption by unintended application of the capability to cause. This important concept of the loss exposure directs the loss prevention effort toward those situations or conditions which have both the probability to occur and a significant potential consequence. For every loss exposure, a probable maximum loss capability can be estimated.

For example, a five-gallon can of gasoline set in the middle of the Sahara Desert has the *capability to cause* a loss, but this loss would not be significant because even if the loss incident (spillage and ignition of the gasoline) occurs, the loss is limited. However, the same amount of gasoline in the basement of a crowded hotel presents a loss exposure that has the *capability* both *to cause* and *to sustain* a loss of major consequence—the hotel and the lives of the occupants.

The loss exposure really consists of many causes and effects. A cause proceeds to an effect, which, in turn, becomes a cause for a subsequent effect. As the loss exposure proceeds toward the loss incident, these individual cause-to-effect sequences, called *causal factors*, can be divided into primary and contributory factors. The primary causal factors are those that are the immediate and sometimes the only causes of the loss incident. Oftentimes, investigators of the loss process are content to find only the primary factor and become involved in another accident because they have not completely protected themselves. The contributory causal factors are those present in such sequence that only the primary causal factor is needed for the process to proceed to the loss incident.

The Loss Incident

The loss incident—that is, the accident itself—is a sudden, undesired event that can result in a loss. The event progresses quickly, going from an incipient condition to full development in a matter of seconds. It has not been expected or predicted. There is a significant loss potential, which will be attained if the process is not arrested, but actually this capability or potential need not be realized for the incident to qualify as a loss incident. A significant loss may occur or some fortunate circumstance may prevent any loss whatsoever. Whether loss occurs or not, the loss incident is absolutely essential if a loss is to occur. The loss incident is a historical fact that, unless its lessons are heeded, is destined to be repeated.

Replacement of a corroded pipe or a worn bearing is not a loss incident. If the pipe corrodes and releases hazardous materials, or if the bearing fails, causing the machine to be hurriedly shut down to prevent damage, these are loss incidents.

Safety engineers are familiar with the "fire triangle." The concept of the three essential elements—the capability to cause a loss, the capability to sustain a loss, and the loss incident itself—form the sides of the "loss triangle." The elimination of any one of the three elements is sufficient to eliminate the loss.

The Loss

Therefore, the products of the loss incident are losses or near-losses. They may be manifest as damage to physical property, injury to people, or an interruption of the manufacturing process.

A loss incident, whether the severity is great or small, does not per se alter the causes or their nature. The severity or the potential severity of the loss calls attention to the loss incident and its causes. Loss incidents with lower degrees of severity cannot be treated lightly, because there is a probability relationship between those incidents and incidents with higher degrees of severity. By reducing the frequency of the minor losses and the near-misses, we are reducing the probabilities of more serious incidents. But when we ignore the loss incidents because only a small loss is experienced, we are displaying attitudes conducive to more serious losses.

It would seem that the losses might be most effectively controlled by the simple prevention of the loss incident. This, while true, is often beyond present capabilities. It is prudent to provide defensive control capability that is held in readiness to be activated after the fact of the loss incident. These defensive efforts, provided by a medical department, fire department, maintenance organization, or specifically designed protective devices, are applied to the loss or the results of the loss incident. The preventive effort of management control must be applied to the loss exposure or the causes of the loss incident.

Loss Control

Exhibit 1 shows a diagram of the loss process and the loss triangle. This is the process through which every loss occurs, and the way to control is indicated.

Safety engineers should realize that the loss process includes causes and effects, that measurement uses the data created by the effects, and that control is achieved through the elimination or minimization of the causes.

Exhibit 1
The Loss Process

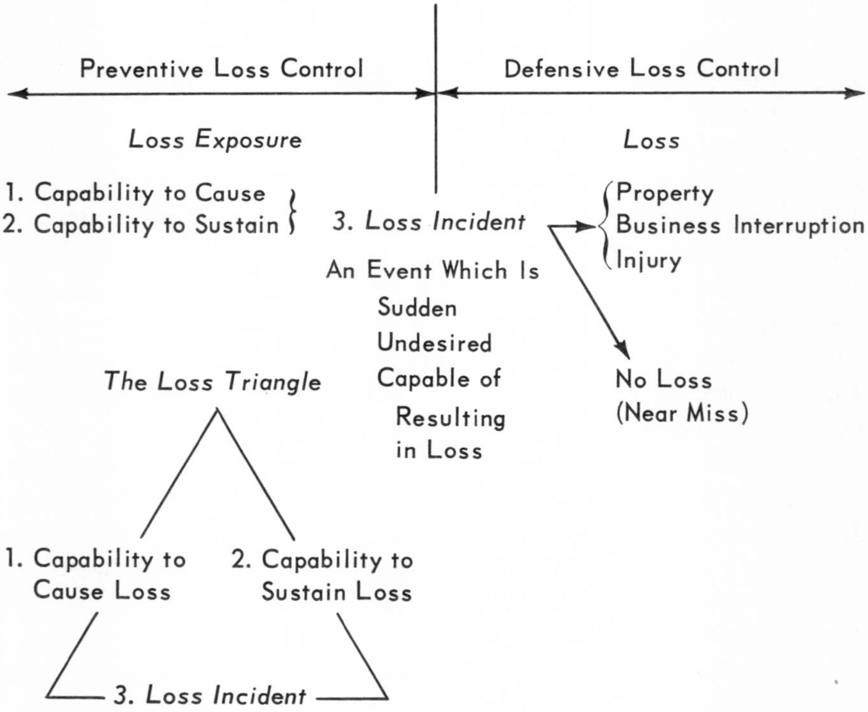

Preventive Loss Control ← | → Defensive Loss Control

Loss Exposure | *Loss*

1. Capability to Cause
2. Capability to Sustain

3. Loss Incident
An Event Which Is
Sudden
Undesired
Capable of
Resulting
in Loss

{ Property
  Business Interruption
  Injury }

No Loss
(Near Miss)

The Loss Triangle

1. Capability to Cause Loss
2. Capability to Sustain Loss
3. Loss Incident

The causes of the incident are the basis for control because, when the causes are recognized and corrected, the loss incident can be thwarted by interrupting the loss process. Practically speaking, loss exposures and incidents cannot be removed from a profit-oriented industry. But we can, at least, reduce the probability and consequence of the loss incident. Loss control, since it requires the authority to act, is a function of the manager and the line supervisor. Every loss incident must have a cause or causes. The idea that most or even some loss incidents can be blamed on bad luck or God does not agree with the facts. Most causes can be identified through searching analysis. The correction is dependent on justifiable and reasonable solutions. For instance, if only a minor injury or a minor property damage resulted, there is no particular motivation to do much of anything. But if the same incident results in an explosion with subsequent fatalities and major property damage, there doesn't seem to be any limit to what management is willing to do to find and correct the causes. This is control justified by the effects, not the causes. An adequate loss prevention program requires that the loss incident effects be treated by the staff people (medical, fire protection, maintenance, and others) and the causes be treated by the line people who have the responsibility, authority, and accountability to act. Logic and reason must overrule emotions. We don't have to wait until someone loses an eye before we put in a mandatory eye protection program or for someone to blow up a boiler before we place standard control devices on boilers. The losses do not generate the causal factors, they only emphasize the problem. The loss incident provides the basis for measurement. The actual losses are known because of their dramatic effects; they can be reported through an accepted routine procedure. The near-misses are also products of the loss incident but are less evident because of the lack of physical damage to people and facilities. Methods for obtaining these reports have been less successful. The results of sampling are also data for measurement, as are the conclusions of systems analysis when the potential loss incidents are developed and analyzed for causal factors. All products of the loss and potential loss incident investigation lend themselves to the development of measurement methods.

In summary, the loss process begins with a loss exposure, which has the capabilities to cause and to sustain a loss. The primary and contributory causes lead to a loss incident, which, in turn, may or may not result in a loss. Through the information obtained from actual and potential loss incidents, measurement of performance is possible; and through analysis of loss exposures, control of performance can be attained.

CHAPTER FIVE

THE PERFORMANCE MEASURING SYSTEM

LET us now focus attention on the measurement area of the management control system. Certain axioms are basic to the accident control system. These axioms may appear to be restraints on the effort to prevent accidents, but, in reality, they provide a practical guide to the measurer and controller of performance. Thomas H. Rockwell has stated certain of them in *ASSE Journal:*

> Elimination of all accidents is an impossible goal regardless of the degree of emphasis.
>
> To err is human. As long as human behavior plays an important role in accident causation, we must admit to the fact that workers will engage in unsafe acts.
>
> Fire fighting or the emphasis of accident prevention efforts on the basis of isolated accidents is an extravagant waste of our resources.
>
> The resources available to combat accidents, namely time and dollars, are not unlimited.
>
> Although the systems approach does not neglect the individual worker, it places emphasis on the aggregate performance.

Accidents are essentially probabilistic in occurrence and in severity particularly. Hence, analysis must be based on a sound foundation of probability theory and statistics.[1]

Every measurement system must meet certain basic criteria. If it is to be effective, a performance measurement system must have the following characteristics.

1. It must be based on a sincere desire to learn the truth so that a factual evaluation is possible.
2. It must accumulate data in significant amounts and in such a way that they can be carefully classified and easily obtained by routine methods.
3. It must analyze these data in a report form that is easily understood by the manager or supervisor of the exposed people or facility.
4. It must be sensitive to change so that deviations can be recognized in time for correction to be effective.

This is the type of performance system that management needs for the control of quantity, quality, and costs. It permits management to do a better job in making decisions and achieving control. Managers want and deserve a comparable method for judging the performance of the loss prevention effort.

The Loss Incident as the Base for Action

The measurement system begins with the loss incident, which marks the occurrence and is a fact. Knowledge of the cause of the loss incident is meaningful only when it becomes the basis for preventive effort. This opportunity to find causes and make corrections is one of the greatest challenges the industrial supervisor has. His ability to control his people and his facilities is the basis for his own supervisor's appraisal of his performance.

There are, of course, various types of losses. The most widely discussed type involves injury to people. The most dramatic type of

---

[1] "A Systems Approach to Maximizing Safety Effectiveness," *ASSE Journal*, December 1961, Vol. VI, No. 6, p. 18.

loss is the one that involves the loss of property usually from fire, explosion, or mechanical breakdown. Probably the most costly type of loss is the business interruption in which not only the production capability is lost while many of the normal operating costs continue to mount up but the position of the product in the marketplace may be lost perhaps long after production can be resumed. But the most frequent type of loss is the one in which no injury, property damage, or business interruption of any actual significance has been suffered. This aborted loss result of the loss incident is fertile ground for causal analysis because many such events occur, but there remains the difficulty of obtaining reports on these events.

Both actual and potential losses are classified by severity. *Minor* losses arbitrarily include a minor injury or a property damage of less than $500. A *serious* loss includes a serious injury or property damage and business interruption of more than $500 but less than $25,000. The *major* loss includes a fatality or disabling injury or property damage and business interruption of more than $25,000. The intention of this classification is not to equate human life with dollars but rather to set arbitrary classifications so that severity analysis is possible.

Two other classifications are possible: First there is the *no-loss*, in which all of the so-called near-misses are included. No-losses may, for measurement purposes, be assigned into any of the other classifications according to their reasonable potential. This reasonable potential must be carefully appraised so that the importance of the loss incident and the effort to find the causal factors are not out of proportion to the emphasis and priority given to other known losses. Because few no-losses are usually reported, the statistical significance of these may be ignored. But when they are reported, the causal factors should be just as important as in any of the other three classifications.

The remaining classification could be termed *disastrous* in that it indicates a loss of more than one life or more than $1 million in property damage and business interruption. Fortunately, these are so infrequent that they are statistically insignificant for measurement as a separate group and can be included in the major classification without changing the significance of this group.

This severity index is an arbitrary classification, and the reader, for reasons of his own, may wish to change the various limits or expand the schedule. In any case, the safety engineer should develop and use a severity index. If he wants valid measurement, he cannot compromise or rationalize the measurement indexes.

Both the type of loss and the severity classifications describe the effect of the loss incident and have little or nothing to do with the causal factors that led to the incident. On the other hand, it is the information gained from the investigation of the loss incident that is vital to the determination of the causal factors and preventative planning.

However, the type and severity classifications are invaluable for two reasons. First, the potential of the loss incident can be appraised. Through simple analysis, we can determine what the reasonable potential could have been had not the defensive mechanisms performed as they did. Such an appraisal can determine the emphasis needed to gain an investigation at the proper level of management and justify the appropriate corrective action. It is important that the appraised potential be "reasonable." It should be the logical, unemotional evaluation of what loss could have been expected, not the assumption that all injury losses would have been fatalities or all property damage losses would have been disasters. Therefore, for measurement purposes, in those loss incidents from which actual losses result, it is more scientific to use the actual loss severity. For investigation, the reasonable potential loss is useful; for measurement, the actual loss is more valuable.

Second, the type and severity classifications are valuable because of the probabilities attendant to the severity classifications in each type of loss. When a certain number of minor injuries are experienced by the workers in an industrial plant, a fraction of these would be serious and a fraction of these, in turn, would be major. This is borne out in Heinrich's much-quoted 300-29-1 ratio of no injury, minor injury, and major injury losses.[2] Monsanto's Texas City plant records show that for the period of 1952 through 1966, there were 270 injuries of minor severity and 29 of serious severity for each major injury. Although this ratio did vary slightly from year to year, the probability relationship remained. Similarly, this probability relationship among the severity classifications of property damage losses also exists. Insurance companies covering prop-

---

[2] H. W. Heinrich, *Industrial Accident Prevention* (New York: McGraw-Hill Book Company, Third Edition, 1953), p. 24.

erty losses and business interruption claims have determined from their years of records that there are probability relationships between minor and serious losses and the major, large, but infrequent, loss. Frank E. Bird, Jr., and George L. Germain found in their studies of 90,000 accidents over a seven-year period that in a group of 500 property damage losses, 300 would amount to less than $50, 150 would cost between $50 and $300, 50 would amount to more than $300, and 1 would be a loss in excess of $1,000. They also showed a cross-probability relationship of 500 property damage losses to 100 minor injury and 1 major injury losses.[3]

The Probability Relationship

This does not mean that each company or plant experiences this same numerical probability. This probability relationship will differ among industries, among companies within an industry, among plants within a company, and among work groups within a plant. But a probability relationship of loss severities will always exist and be useful in the measurement of performance.

There is an advantage in the fact that the relationship does vary as just described. A plant's performance cannot really be compared to that of some other plant; it can show improvement only as it improves its own experience. Meaningful measurement depends on the sincere desire to learn the truth and is hindered by the desire to appear safer than someone else in a contest situation. A plant having a major loss for every 100 losses is not necessarily more or less safe than a plant which experiences a major loss for every 200 losses. It is true, however, that if both plants reduced the frequency of their loss experience, they would have reduced the probability of the major loss and, therefore, improved their performance.

The Loss Report

In order to obtain the information needed for the measurement system, there must be a report of the loss incident. This report must

[3] Frank E. Bird, Jr., and George L. Germain, *Damage Control* (New York: American Management Association, Inc., 1966), p. 45.

mark the occurrence and relate exactly what happened with what result.

When there is an actual loss, the report may be triggered by the loss incident. Because the loss is evident, the reporter can report the needed information, and those who are responsible for the loss can perform the appropriate investigation into the causal factors. The procedure should be simple so that reporters don't become discouraged. The report form should be designed so that the needed information is specifically requested and recorded in a manner that easily lends itself to analysis. The report form should permit the transfer of the loss incident information to cards or tapes for computer use.

When the output is a potential rather than an actual loss, the report can be the product of several different sources, depending on the circumstances. It can be written by an individual who has been involved in or observes a near-miss loss incident. The report can also be the product of a program or method, such as sampling. Or it may be the result of research or design work. In other cases, a person may recognize a potential hazard in a certain process or piece of equipment.

Potential losses are reported much less often than are actual losses. For example, if a scaffold board falls from the third deck, the actual loss could be a fatality to an individual struck by the board. The potential loss involves the same fatality, but if no one was hurt, the accident of the falling board might never be reported. Because of this problem, the two reports (if the potential loss *is* reported) should not be combined for the measurement method but rather should be treated separately.

There is also a difference in the reporting of injury losses and of property damage or business interruption losses. An injured individual can either go or be taken to a place where medical care is available. In this way, the injury loss reports itself. The damaged equipment cannot take itself to the shop for the needed repairs, so the report of the loss incident is dependent on the owner or some other individual.

This report may be a voluntary incident report by the owner or other responsible person, a work order to repair the damaged equipment, or a routine production report showing a loss of production time or a lowered output. Unless there is a routine, accepted procedure by which reports are sent to the measurer, he may never know that the incident has occurred. This is especially true in the case of near-miss or potential losses.

Psychological Problems of Reporting

Attitudes greatly affect the reporting of loss incidents. On each level of supervisory management, there is a reluctance to report any event that might reflect adversely upon the supervisory performance. The obvious loss will be reported because it cannot be hidden, but the minor losses, which could be the basis for further loss prevention, go unmentioned. Managers want reporting but tend to throttle it by implying that the supervisory group that reports losses must be performing poorly. An example of this would occur when a corporate official asks his division managers to report all losses over $5,000 to him. Suppose that after one quarter he receives a report that Division A had no losses, Division B had eight losses totaling $80,000, and Division C had one loss costing $200,000. The corporate official shoots a letter off to the manager of Division B asking why he had so many losses and stating that corporate experts would be assigned to help solve his problem if improvement is not immediately apparent. The next quarterly report will show the expected improvement—no losses from Division B. Of course, losses continued to occur in all divisions in both quarters, but the desire for high performance overshadowed the desire for loss prevention and control. Managers A and C knew the rules of the ball game (C just couldn't hide the large loss), and Manager B soon learned.

Lower levels of supervision also want reporting but without the need of reporting, in return, to anyone higher in the organization. They also want assistance from staff groups, but prefer that this activity proceed without comment to the boss. Workers have personal feelings about reporting, too. Some would report anything in order to gain a kind of recognition or, in the case of a very minor injury, just to get some time away from the job. Others avoid reporting incidents because they feel that the reporting is not their responsibility or that they will be blamed for the incident. If an organization applied no explicit form of pressure toward the incident or the report, reporting would still lag, because every responsible human being has a conscience, and a loss incident poses an implied failure of performance. However, this pressure within man himself also drives him to perform well.

But the safety engineer needs the report for measurement purposes. He must be dedicated to this need in order to fulfill his responsibility. He must develop integrity and respect for his office. If necessary, he must provide a shield for the report and the reporter while, at the same time, giving an honest and frank appraisal of performance. Although it is impossible to avoid the pressure, either explicit or implied, associated with reporting, it is possible to win enough support on all levels of supervisory management so that an acceptable level of reporting is obtained.

Reporting must be encouraged or mandated so that an adequate base for measurement and appraisal is provided. Reporting must be sold on its positive features. The safety engineer should recognize the reasons for the report and the measurement system and convey these reasons as positive features of the management control program. The reports are vital since they provide the basis for effective control through measurement and investigation of causal factors. Control is the purpose; measurement is the means. The report does record the loss incident but this is secondary to the opportunity it affords for investigation and correction. Performance should be evaluated according to how well losses are prevented, not how few losses are reported. If losses are not recognized, both specifically and as trends, the ultimate performance will be poorer and the losses more severe.

Complete Coverage Necessary

Reports from only one source of losses or including only one type or severity of loss are not sufficient. It has been both traditional and necessary to place emphasis on loss incidents which have or could have resulted in personal injuries. However, this represents only a part of the loss prevention problem; it ignores the areas of property damage and business interruption.

The report of the loss incident is not a report of performance—it is a report of an event. Of course, the supervisor is sorry, perhaps ashamed, that the event has occurred. After an appropriate period of self-pity, he must recognize that the event *has* occurred and that nothing that he does now can turn the clock back to avoid that event. But he can do something about the next one, the similar one that may happen across the street in Unit B or at Company Y. The report can do so much to help—to convey needed information, to classify this event among other similar ones, to alert other talented people for their assistance, to initiate the determination of cause and seek correction, and, finally, to provide a base for measurement. From this activity, the ability to manage is demonstrated, and performance can be appraised for appropriate recognition. Both manager and supervisor must accept this. Rattling of sabres or poison-pen memorandums will not stop the accidents— only the accident reports.

The detection of the loss incident that has not occurred *yet* is even more rewarding because a possible loss may be avoided. All the advantages of the actual loss incident report are gained without an actual loss.

The report form is also important. The report must fit the situation and the people—that is, the local needs. The form should be simple yet complete; only needed information and facts should be included. The reporter should be aware of pressures and respect confidences.

It should be emphasized that the purpose of measurement is to gain better control so that losses are prevented and profits protected. The measurement program should analyze all available data as a unit. It should be based on the actual incidents that involve losses by injury, property damage, and business interruption. It should seek potential loss incidents by sampling and by process or mechanical loss analysis. It should provide an easily understood report to management that compares present performance with past performances, indicates trends, and recommends priority of emphasis. The objective measurement system should guide control without crisis and motivate action without reprisal.

In the following chapters, each of the components of the measurement system will be described. The technique used is neither unique nor is it necessarily the best, but it has been used effectively. It should challenge the safety engineer, whose responsibility is the evaluation of performance in loss prevention.

CHAPTER SIX

THE INJURY REPORT TO CONTROL

THE first area of available data that can be measured is the accident that results in injury. The traumatic effects are the very reason that these data are available. The value of the data depends on the employee–employer attitudes about reporting and care received, the method of recording and transmitting the information, the technique of handling the data, and finally the use of the recommendations resulting from the data analysis.

Employee Attitudes

The report of the accident is initiated because the person has sustained an injury and seeks some type of assistance or treatment to relieve the physical or financial suffering. The employee can be motivated to report the injury for several reasons. He may require or desire medical attention. He may accept (or be fearful of the consequences of disregarding) the management policy, rule, or procedure that tells him to report "any injury no matter how minor." He may wish to establish his claim for financial coverage for the medical cost and the loss of wages that might result from the injury. The employer should encourage this reporting, thereby demonstrating humanitarian concern for the employee and obtaining, in the long run, the least cost for the injury.

Whatever the motivation may be, the injured appears at the medical station, where he receives treatment and can provide information about the accident. Therefore, it is logical that the medical station personnel act as the recorder of the incident. This contact between the medical people and the injured is very important to the accident reporting and investigation process. It is in this contact that the employer's interest in and concern for the well-being of the employee is most personally demonstrated. The attitude of the employee toward the employer, not only in matters of safety but in many other matters of personal concern, will be affected or even fixed by the words and actions of the medical group. Care for the injury is the first order of business, and this must be competently and promptly given.

Reporting or the recording of information is second but not secondary. It is not the responsibility of the medical personnel to evaluate the reasons for the accident or to pass judgment on the validity of the injured's comments concerning conditions or causes leading to the incident (even though they often must listen to these). It is their responsibility to provide proper medical treatment and record the information required by the reporting procedure.

The Value of Data

The information in the report can serve the safety program in many ways. It pinpoints the occurrence as to when, where, and how the accident occurred so that full investigation can follow. It establishes the injured's claim to medical and wage benefits provided by law, work contract, or company supplemental benefit program. And, most important, it provides an opportunity for an interested and responsible supervisor to demonstrate his concern for his people through investigation of the accident and correction of the causes. Probably at no other time is an employee's attitude toward his job, his supervisor, and his company so affected as it is when he is injured. And the attitude of the medical personnel and the supervisor toward the injury and the accident will greatly affect the performance measuring system's functioning.

The supervisor's method of investigation also affects employee attitudes toward safety and injury reporting. This is a difficult time for both injured and supervisor. The injured feels that there will be some embarrassing questions to be answered and that he will, in all probability, be blamed for the incident. He knows that the supervisor is under pressure to prevent incidents like this in his area and that the supervisor needs to save face if possible. The injured would prefer to avoid the investigation; therefore, he may not report injuries—at least the minor ones.

The supervisor also approaches the investigation with some apprehension. He recognizes that his performance rating is threatened, and he has other matters that need his attention. However, loss investigation is one of the most important responsibilities of the supervisor. In this activity, he has the opportunity to show his concern for his employees and to demonstrate his safety standards. It is the supervisor's attitude that ultimately determines the extent to which employees are willing to make reports of incidents. Without a responsible attitude for safety on the part of the supervisor, the reporting of incidents and, thus, the opportunity to correct causes is lost. The true performance of the supervisor is not a function of the number of incidents that occur in his area but of his effectiveness in recognizing and correcting causes.

Having shown his interest in the employee by being sure that the medical treatment has been given for the injury, the supervisor seeks the facts about the incident, develops the causes that led to this incident with no intent to find a "goat," and plans with the injured what action will be appropriate to minimize or eliminate the specific causes. This businesslike, logical process of problem solving results in an effort commensurate with the actual and potential result of such an incident. This positive approach gains employee respect for both the supervisor and the program and encourages prompt reporting of incidents.

The supervisor has an additional responsibility and opportunity to share this incident and its lessons with other supervisors who have similar hazards. Once an accident has occured, it would be a shame to have every supervisor and employee learn the same lessons by bitter, possibly more serious experience. Like the action to be taken to minimize the causes of this accident, the effort to reach other supervisors should be in relation to both the actual and the potential results of such incidents.

The Report Form

The report form, on which the pertinent facts obtained by the medical personnel and the results of the investigation are recorded, should be sent to the safety engineer. Here, the information can be checked for omissions and corrected if necessary before being entered into the measuring system. Depending on how the data are handled, the extent of the analysis, and the methods for communicating the conclusions, the report form may differ somewhat.

A report form of the type shown in Exhibit 2 fulfills most of the criteria for analysis. First, it gives a complete picture of the accident. It identifies the injured, places the time and location of the accident, describes the extent of the injury and the part of body involved, and fixes the agency, type, and cause of the accident. The statement of the injured and the corrective action planned by the supervisor are recorded by written statements.

Second, the form is easy to fill out. For the most part, the notations are made by circling appropriate numbers, and very little writing is required. This is important because the acceptance of the reporting process and, often, the investigation process depends on how much time and effort are required. This is especially true when the injury is very minor and the potential of the accident is limited. If the accident deserves an extensive investigation with more formal reporting to various organizational levels, this can be done. In that case, the short form records the incident, while the formal report discusses the findings of the investigation.

Third, the information of the report can be tabulated easily or keypunched onto cards. The use of numbers to denote certain pieces and groups of information provides great versatility and scope to the business of analysis. With computers or manually tabulated lists, accidents can be grouped by cause, injury, scene, and occupation. Having the information in a form that is easy to tabulate makes it easy to present this information to management quickly, upon demand if necessary.

Exhibit 2

Accident and Injury Report Form

PLEASE DO NOT WRITE IN THIS SPACE

ACCIDENT AND INJURY REPORT

→ THIS FORM MUST BE COMPLETED AND RETURNED TO THE SAFETY DIRECTOR WITHOUT DELAY AFTER DISPENSARY TREATMENT

1- Date -6 Of Accident	7- Time -10 Of Accident	Sex 21	Overtime 22	26- Name Of Injured Person
03/15/49	0930	①Male 2 Female	1 Yes ②No	John Doe -40

42- Badge No. -45	Day of Week 49		Supervisor
2941	1 Mon. 2 Tues. 3 Wed 4 Thurs. 5 Fri.⑥Sat. 7 Sun.		FRANK SMITH

EMPLOYEES STATEMENT AS TO CAUSE OF INJURY

I was closing a steam valve and hit my arm on an uninsulated steam line.

TREATMENT AND COMMENT

Applied burn ointment and dressing.

Treated By Jane Brown, R.N.

INJURED PART (Circle One or Two)

CC 51-53, 54-56
100 Skull
110 Scalp
120 Face
13 _ Ear*
14 _ Eyes*
210 Nose-Throat
200 Neck
30 _ Shoulders
310 Back
320 Chest
400 Abdomen
41 _ Inguinal*
420 Rectum
430 Genitalia
150 Teeth

*For These Parts, Indicate: 1-Right, 2-Left or 3-Both

Upper Extremities Lower Extremities
50 _ Upper Arm* 60 _ Thigh*
51 _ Elbow* 61 _ Knee*
52⑦Lower Arm* 62 _ Leg*
53 _ Wrist* 63 _ Ankle*
54 _ Hand* 64 _ Foot*
55 _ Thumb* 65 _ Great Toe
56 _ 2nd Finger* 66 _ Any Other Toe*
57 _ 3rd Finger* _ _ Multiple
58 _ 4th Finger* 70 _ Upper body*
59 _ 5th Finger* 71 _ Lower Body*

TYPE OF INJURY (Circle One)

CC 57-58
01 Laceration 11 Fract.-Comp.
02 Incised Wnd. 12 Contusions
03 Puncture ⑬Burn
04 Amputation 14 Irritation
05 Avulsion 17 Inhalation of fumes
06 Abrasion
07 Sprain 18 Internal
08 Strain 19 N.A.I.
09 Dislocation 20 Blister
10 Fract. Simple

SEVERITY (Circle One)

CC 59
①Minor 2 Dr. Case 3 Fatal
4 Major 5 Serious
6 Non-Industrial

CC 72, 73, 74
HOME DEPARTMENT
016

WEATHER (Circle One)

CC 75
①Clear 4 Hail 6 Fog, Smoke, or Haze 9 Dry Norther
2 Rain 5 Freezing Rain 7 Thunderstorm
3 Drizzle Snow or Sleet 8 Hurricane

SCENE OF ACCIDENT (Circle One)

CC 76-77
01 Dept. 1 14 Dept. 19 28 Research 41 Salvage
02 Dept. 2 15 Dept. 20 29 Control Lab 42 Streets
03 Dept. 3 17 Dept. 22 30 P. E. Lab 43 Yards and Walks
04 Dept. 4 18 Dept. 23 31 Stores 44 North 80
05 Dept. 5 19 Dept. 24 32 Shops 45 North 68
06 20 Dept. 25 33 Instr. Shop 46 Locker Bldgs.
07 Dept. 7 21 Dept. 26 34 Docks 47 Emp. & Guard Offices
08 Dept. 8 22 Dept. 31 35 Truck Racks 48 Misc. Maint. Areas
09 Dept. 10 23 Dept. 41 36 Tank Car Racks 49 Pilot Plant
⑩Dept. 11 24 Dept. 42 37 Power 1 50 Product App. Bldg.
11 Dept. 6 25 Dept. 43 38 Power 2 51 Pipe Racks
12 Dept. 17 & 18C 26 Main Office 39 Power 3 52 Warehouses
13 Dept. 18-P 27 Office Annex 40 Utilities—Other 53 Dept. 33
         54 Dept. 34
         55 Dept. 35

OCCUPATION (Circle One)

CC 78-79
03 Analyst 52 Ironworker 04 Pumper Guager 89 Clerical
50 Asbestos Worker 30 Janitor 57 Sheet Metal 91 Exempt Proc. Tech.
51 Boilermaker 60 Laborer 62 Trucker 92 Exempt-Maint.
20 Cafeteria 56 Machinist 65 Warehouseman & Stores Clerk 93
53 Carpenter ⑥Oper. 94 Exempt-Operating
54 Electrician 64 Oper. Engr. 71 Engr. Aide 95 Exempt-Other
10 Guard 55 Painter 79 Lab Tech. 99 Exempt-Res.
63 Inst. Man 58 Pipefitter 81 Non-Exempt-Other

(FRONT)

Exhibit 2
Accident and Injury Report Form (continued)

AGENCIES (Circle Only One)

THE OBJECT, SUBSTANCE OR EXPOSURE MOST CLOSELY ASSOCIATED WITH THE INJURY

MACHINES
- 0000 Agitator, Mixer
- 0400 Conveyor
- 0017 Drill
- 0009 Extruder
- 0005 Floor Polisher
- 0018 Lathe
- 0015 Mill
- 0053 Mower
- 0021 Packaging Mach
- 0083 Picture Proj.
- 0086 Pile Driver
- 0094 Pipe Threader
- 0026 Planer, Shaper
- 0042 Saw (Table)
- 0048 Shears & Dicer
- 0038 Rolls
- 0090 Sandblast Mach.
- 0004 Sander, Grinder
- 0044 Hacksaw (Power)
- (0052) Jointer
- 0029 Presses
- 0007 Valve Grinder
- 0012 Welding Machine

STEAM, BOILERS and PRESSURE VESSELS
- 1510 Boilers
- 0523 Columns
- 0521 Exchangers
- 1931 Filters
- 0074 Furnace
- 0530 Hose
- (0525) Piping
- 0528 Tanks
- 0529 Trap
- 0522 Reactor

ELECTRICAL APPARATUS
- 0940 Batteries
- 0920 Conductors
- 0902 Generators
- 0960 Heating Appl.
- 0901 Motors
- 0930 Switch Gear
- 0910 Transformers

PUMPS AND PRIME MOVERS
- 0134 Compressors
- 0110 Eng. & Prime Movers
- 0124 Fans & Blowers
- 0120 Pumps

FLAMMABLE and HOT SUBSTANCES
- 1225 Fire
- 1954 Metal Plate
- 1255 Steam
- 1956 Water
- 1953 Welding Rod or Slag

HAND TOOLS
- 1010 Axe
- 1011 Blowtorch
- 1012 Broom, Brush
- 1013 Chisel, Punch
- 1014 Dies & Taps
- 1015 Drill (Hand)
- 1053 Drill (Power)
- 1016 File
- 1019 Pry Bar, Cheater
- 1017 Glasscutter
- 1051 Grinder
- 1018 Hammer, Hatchet
- 1059 Hydroblaster
- 1020 Jack
- 1021 Knife
- 1025 Oilcan
- 1026 Pick, Fork, Pickaxe
- 1023 Pipecutter
- 1028 Plane
- 1027 Pliers
- 1056 Pneumatic Hammer
- 1062 Torch
- 1065 Sandblaster
- 1032 Sandpaper
- 1030 Saw (Hand)
- 1066 Saw (Power)
- 1069 Screwdriver
- 1032 Shovel, Spade
- 1031 Sickle
- 1029 Tapeline
- 1070 Torch
- 1035 Wrench

CHEMICALS, DUST and RADIATION
- 1121 Acid
- 1157 Additives
- 1132 Ammonia
- 1128 AN
- 1134 Benzene
- 1122 Carbon Monox.
- 1135 CCl 4
- 1158 Catalyst
- 1133 Caustic
- 1123 Chlorine
- 1159 DM
- 1300 Dust
- 1152 EB
- 1153 EDC
- 1110 Explosive Gases
- 1124 HCN
- 1129 Hexane
- 1131 Lime
- 1125 Methanol
- 1126 Mercury
- 1155 Nitrogen
- 1137 Oils
- 1138 Polymer
- 1235 Paint, Varnish
- 1400 Radiation
- 1127 Refrigerants
- 1156 Solvents
- 1151 Styrene
- 1140 Sulphur
- 1153 VAM
- 1154 VCM
- 1160 Unknown

MECHANICAL POWER TRANSMISSION
- 0814 Belts
- 0812 Chain
- 0804 Couplings
- 0810 Drums
- 0807 Gears
- 0809 Pulleys
- 0811 Ropes, Cables

ELEVATORS and HOISTING APPARATUS
- 0310 Crane
- 0200 Elevators
- 0330 Hoists
- 0320 Shovel

WORKING SURFACES
- 1530 Floors
- 1578 Roads
- 1576 Roofs
- 1583 Sidewalks
- 1582 Stairs, Ramps

VEHICLES
- 0650 Aircraft
- 0640 Barges, Ships
- 0660 Bicycles
- 0611 Fork-lift
- 0691 Hand trucks, Tool Box & Wheelbarrows
- 0610 Motor Vehicle
- 0630 Railway

MISCELLANEOUS
- 1916 Bags, Boxes
- 1914 Barrels, Drums
- 1915 Bottles (ICC)
- 1918 Cans
- 1923 Ditches, Trenches
- 1922 Doors, Windows
- 1930 Floor Openings
- 1932 Food
- 1934 Glassware
- 1979 Handrails
- 0720 Insects & Snakes
- 1947 Kitchen Equip.
- 1952 Lab Equip. Misc.
- 1950 Ladders
- 1951 Load. Platforms
- 1954 Manholes
- 1958 Nails
- 1957 Nuts, Bolts
- 1963 Office Equip. & Machines
- 1966 Person
- 1967 Pipefittings & Valves
- 1978 Scaffold
- 1980 Splinter, Silver
- 1581 Structure or Brace
- 1991 Wire-Non-Elect.
- 1992 Workbench, Desk

1999 Misc. Matl. or Obj. _____ Please Specify

ACCIDENT TYPE (Circle One)
CC 65-66
- 00 Striking Against
- 01 Struck By
- 02 Caught in or Between
- 03 Fall - Same level
- 04 Fall - Different level
- 05 Slip
- (06) Temperature Extreme
- 07 Inhalation, Ingestion
- 08 Electrical Contact
- 09 Chemical Contact
- 10 Lifting, Pushing, Pulling
- 20 N.E.C.

UNSAFE MECHANICAL OR PHYSICAL CONDITION (Circle One)
CC 67
- (0) Improperly Guarded
- 1 Defect of Agency
- 2 Housekeeping
- 3 Improper Illumination
- 4 Improper Ventilation
- 5 Defective Apparel
- 6 Improper Design
- 9 (Please Specify)

UNSAFE ACT (Circle One)
CC 68-69
- 00 Operating Without Authority
- 01 Using Unsafe Speed
- 02 Making Safety Devices Inoperative
- 03 Using Unsafe Equipment or Using Unsafely
- 04 Unsafe Loading, Piling, Mixing
- (05) Unsafe Posture or Position
- 06 Working on Moving Equip.
- 07 Horseplay, etc.
- 08 Failure to Use Protective Equipment or Clothing
- 09 (Please Specify)
- 10 Improper Method
- 11 Failure to be Warned

UNSAFE PERSONAL FACTOR (Circle One)
CC 70
- 0 Improper Attitude
- 1 Lack of Knowledge or Skill
- 2 Bodily Defects
- 3 Lack of Instructions
- 4 Inattention - Injured
- 5 Willful Disregard of Safety Rule or Instruction
- 6 Inexperience
- 7 Inattention - Fellow Worker
- (9) None (Please Specify)

Is this a late report? CC 71 3 Yes (4) No
If "Yes", please comment below.

Correction Action Taken To Prevent Recurrence

Employee was cautioned to use extreme care when working in close quarters. Work order has been written to insulate the pipe.

Frank Smith
(Supervisor's Signature)

(BACK)

To use this form, the medical personnel, having treated the injured, complete the front side of the report form. This gives the pertinent personal and time information, the type and severity of the injury, the scene of the incident, and the occupation of the injured. This information will provide the answers to such questions as: What work groups are experiencing the incidents? When and where are the incidents occurring?

The report form should then be sent by the medical personnel or taken by the injured to the responsible supervisor. When the supervisor has completed his investigation of the incident, he records his findings on the back of the form. This provides the "why" of the incident and the statement of what action he intends to take to correct the causal factors.

The report form in Exhibit 2 was prepared and is used by a chemical plant. It fits the needs of that plant but might not fit every one and would probably need considerable revision to be used in a foundry or sawmill. However, the same concept can be used by all industry; only revisions in some of the wording are required.

The USASI Standard Z16.2 gives a general but comprehensive list of items within numbered sequences that can be used in the development of the report form for industrial establishments or whole companies. It is possible to develop a report form (an example is shown in Exhibit 3) that could be used by all industry and collected by the National Safety Council. It would provide a detailed picture of the safety performance by industry, section of the country, or whatever breakdown is desired to whoever desires the information. It is possible, but it probably won't happen. For the same reason that most laymen cannot accept the fact that the highways are safer today than they were ten years ago in spite of the fact that more people are killed in highway accidents and some politicians tenaciously believe that safety can be legislated, this information about industry, which could be of great value in reducing accidents to people, will probably never be available, not even to the National Safety Council. But for the plant—its safety engineer and its manager—the report form in simple style can be a very effective tool for the reduction of accidents that result in injury to employees.

The information derived from the report form can be tabulated

Exhibit 3

General Accident and Injury Report Form

PLEASE DO NOT WRITE IN THIS SPACE
(For Medical Personnel's Use)

ACCIDENT AND INJURY REPORT

42 – Badge – 45	7 – Time – 10 Of Accident	Sex	Overtime	26 – Name of Injured – 40
	1 Mon, 2 Tues, 3 Wed, 4 Thur, 5 Fri, 6 Sat, 7 Sun.	1. Male 2. Female	1. Yes 2. No	Supervisor or Foreman Person

EMPLOYEE'S STATEMENT AS TO CAUSE OF INJURY

TREATMENT AND COMMENT

INJURED PART (Circle one or two)

HEAD
- 110 Brain
- 120 Ear(s)
- 130 Eye(s)
- 140 Face
- 150 Scalp
- 160 Skull
- 198 Head (Multiple)
- 199 Head NEC
- 200 Neck

UPPER EXTREMITIES
- 311 Upper Arm
- 313 Elbow
- 315 Forearm
- 318 Arm (Multiple)
- 320 Wrist
- 330 Hand
- 340 Finger(s)
- 398 Extremities (Multiple)
- 399 Extremities NEC

LOWER EXTREMITIES
- 511 Thigh
- 513 Knee(s)
- 515 Lower Leg
- 518 Leg (Multiple)
- 520 Ankle
- 530 Foot
- 540 Toe(s)
- 598 Extremities (Multiple)
- 599 Extremities NEC

TRUNK
- 410 Abdomen
- 420 Back
- 430 Chest
- 440 Hips
- 450 Shoulder(s)
- 498 Trunk (Multiple)
- 499 Trunk, NEC

BODY SYSTEM
- 801 Circulatory System
- 810 Digestive System
- 820 Excretory System
- 830 Musculo-Skeletal System
- 840 Nervous System
- 850 Respiratory System
- 880 Other Body Parts
- 900 Body Parts, NEC

NATURE OF INJURY (Circle one)

- 100 Amputation
- 110 Asphyxia Drowning, Strangulation
- 120 Burn (heat)
- 130 Burn (chemical)
- 140 Concussion
- 150 Disease
- 160 Contusion
- 170 Cut, Laceration, Puncture
- 180 Dermatitis
- 190 Dislocation
- 200 Electric Shock
- 210 Fracture
- 220 Freezing
- 230 Hearing
- 240 Heat Effects
- 250 Hernia
- 260 Joints, Tendons, Muscles
- 270 Poisoning
- 280 Pneumoconiosis
- 290 Radiation Effects
- 300 Abrasions
- 310 Sprains, Strains
- 400 Multiple Injuries
- 990 Occupational Disease, NEC
- 995 Other Injury, NEC
- 999 Unclassified Not Determined

TREATED BY

SEVERITY (Circle one)
1. Minor 2. Dr. Case 3. Fatal
4. Major 5. Serious 6. Non-Occupational

WEATHER (Circle One)
1. Clear
2. Rain
3. Drizzle
4. Hail
5. Freezing Sleet or Snow
6. Fog, Smoke or Haze
7. Thunderstorm
8. Hurricane or Cyclone
9. Dry Norther
10. NEC

SCENE OF ACCIDENT (Circle One)
- 06 Machine Shop
- 07 Warehouse
- 08 Loading Rack
- 09 Shipping & Receiving
- 10 Office Building
- 11 Streets
- 12 Grounds
- 13 Parking Lot
- 14 Dock
- 15 Laboratory

OCCUPATION OF INJURED (Circle One)
- 001 Craftsman
- 002 Truck Driver
- 003 Machine Operator
- 004 Cook
- 005 Baker
- 006 Laborer
- 007 Inspector
- 008 Stock Clerk
- 009 Janitor
- 010 Technician
- 011 Engineer
- 012 Foreman
- 013 Secretary
- 014 Clerical
- 015 Manager

- 01 Department 1
- 02 Department 2
- 03 Department 3
- 04 Department 4
- 05 Department 5

(FRONT)

Exhibit 3
General Accident and Injury Report Form (continued)

SOURCE OF INJURY* (Circle One)
*Name as the source of injury the object, substance or bodily motion which directly produced the injury previously identified in the nature of injury classification.

- 0100 Air Pressure (Abnormal)
- 0200 Animals, Insects, Birds, Reptiles
- 0300 Animal Products (Not Food)
- 0400 Bodily Motions (Not Lifting)
- 0500 Boilers, Pressure Vessels
- 0600 Boxes, Barrels, Containers
- 0700 Building and Structures
- 0800 Ceramic Items, NEC
- 0900 Chemicals & Compounds
- 1000 Clothing, Apparel, Shoes
- 1100 Coal and Petroleum Products
- 1200 Cold (Atmospheric)
- 1300 Conveyers
- 1400 Drugs & Medicines
- 1500 Electric Apparatus
- 1600 Excavations
- 1700 Flame, Fire & Smoke
- 1800 Food Products
- 1900 Furniture, Fixtures
- 2000 Glass Items, NEC
- 2100 Hand Tools (Not Powered)
- 2200 Hand Tools (Powered)
- 2300 Heat (Atmospheric)
- 2500 Heating Equipment, NEC
- 2600 Hoisting Apparatus
- 2700 Infectious & Parasitic Agents, NEC
- 2800 Ladders (Fixed or Portable)
- 2900 Liquids, NEC
- 3000 Machines
- 4000 Mechanical Power Transmission Apparatus
- 4100 Metal Items, NEC (Plates, Rod, Wire, etc.)
- 4200 Mineral Items, Metallic, NEC
- 4300 Mineral Items, Non-Metallic, NEC
- 4400 Noise
- 4500 Paper and Pulp Items, NEC
- 4600 Particles (Unidentified)
- 4700 Plants, Trees, Vegetation
- 4800 Plastic Items, NEC
- 4900 Pumps and Prime Movers
- 5000 Radiation Substances and Equipment
- 5100 Soaps, Detergents, Cleaning Compounds, NEC
- 5200 Silica
- 5300 Scrap, Debris, Waste Material, NEC
- 5400 Steam
- 5500 Textile Items, NEC
- 5600 Vehicles
- 5700 Wood Items, NEC
- 5800 Working Surfaces
- 8800 Miscellaneous, NEC
- 9800 Unknown, Unidentified

ACCIDENT TYPE (Circle One)
- 010 Struck Against
- 020 Struck By
- 030 Fall From Elevation
- 050 Fall On Same Level
- 060 Caught In, Under, Between
- 080 Rubbed or Abraded
- 100 Bodily Reaction
- 120 Overexertion
- 130 Electric Current
- 150 Temperature Extremes
- 180 Chemical Contact or Radiation
- 200 Public Transportation
- 300 Motor Vehicle Accidents
- 899 Accident Type, NEC
- 999 Unclassified, Insufficient Data

HAZARDOUS CONDITION (Circle One)
- 000 Defect of Agency
- 100 Apparel Hazards
- 200 Environmental Hazards
- 300 Hazardous Procedures
- 400 Placement Hazards, Housekeeping
- 500 Inadequately Guarded
- 600 Outside Workplace Hazards
- 700 Public Hazards
- 980 Hazardous Conditions, NEC
- 990 Undetermined
- 999 No Hazard Condition

AGENCY OF ACCIDENT* (Circle One)
*Name the object, substance or premises to which the hazardous condition which was noted applies.

- 0100 Air Pressure (Abnormal)
- 0200 Animals, Insects, Birds, Reptiles
- 0300 Animal Products (Not Food)
- 0500 Boilers, Pressure Vessels
- 0600 Boxes, Barrels, Containers
- 0700 Building and Structures
- 0800 Ceramic Items, NEC
- 0900 Chemicals & Compounds
- 1000 Clothing, Apparel, Shoes
- 1100 Coal and Petroleum Products
- 1300 Conveyers
- 1400 Drugs and Medicines
- 1500 Electric Apparatus
- 1600 Excavations
- 1800 Food Products
- 1900 Furniture, Fixtures
- 2000 Glass Items, NEC
- 2200 Hand Tools (Not Powered)
- 2300 Hand Tools (Powered)
- 2500 Heating Equipment, NEC
- 2600 Hoisting Apparatus
- 2700 Infectious & Parasitic Agents, NEC
- 2800 Ladders (Fixed or Portable)
- 2900 Liquids, NEC
- 3000 Machines
- 4000 Mechanical Power Transmission Apparatus
- 4100 Metal Items, NEC (Plates, Rod, Wire, etc.)
- 4200 Mineral Items, Metallic, NEC
- 4300 Mineral Items, Non-Metallic, NEC
- 4500 Paper & Pulp Items, NEC
- 4600 Particles (Unidentified)
- 4700 Plants, Trees, Vegetation
- 4800 Plastic Items, NEC
- 4900 Pumps & Prime Movers
- 5000 Radiating Substances and Equipment
- 5100 Soaps, Detergents, Cleaning Compounds, NEC
- 5300 Scrap, Debris, Waste Materials, NEC
- 5500 Textile Items, NEC
- 5600 Vehicles
- 5700 Wood Items, NEC
- 5800 Working Surfaces
- 5900 Work Area or Environment, NEC
- 8800 Miscellaneous, NEC
- 9800 Unknown, Unidentified
- 9999 No Agency of Accident

AGENCY OF ACCIDENT PART (Circle One)
- 0500 Parts of Boilers and Vessels
- 0700 Parts of Buildings and Structures
- 1300 Parts of Conveyers
- 2200 Parts of Hand Tools Not Powered
- 2300 Parts of Hand Tools, Powered
- 2600 Parts of Hoisting Apparatus
- 3000 Parts of Machines
- 5600 Parts of Vehicles
- 9800 No Agency Part Indicated
- 9900 No Agency Part

UNSAFE ACT CLASSIFICATION (Circle One)
- 050 Working on Moving Equip.
- 100 Failure to Use Protective Equip.
- 150 Failure to Wear Safe Attire
- 200 Failure to Secure or Warn
- 250 Horseplay
- 300 Improper Use of Equip.
- 350 Improper Use of Body Parts
- 400 Inattention
- 450 Making Safety Devices Inoperative
- 500 Working at Unsafe Speed
- 550 Unsafe Position or Posture
- 600 Driving Errors
- 650 Unsafe Loading, Piling, Mixing
- 750 Using Unsafe Equip.
- 900 Unsafe Act NEC
- 998 No Unsafe Act
- 999 Unclassified

CORRECTION ACTION TAKEN TO PREVENT RECURRENCE

_____

_____

_____

_____

(Supervisor's Signature)

(BACK)

by cause, injury, scene of the accident, or occupation of the injured. The information can be presented by graphs or charts to management and supervisory personnel to show the month-to-month story of safety performance. These presentations give a continuous means of measuring the effectiveness of the accident prevention effort and place this information, in an easily readable form, in the hands of management within a few days after the end of the month, just as managers receive monthly figures from cost accounting.

The Control Chart

A representative sample of the control chart is shown in Exhibit 4 to illustrate how this method of using injury statistics can guide the accident prevention effort. First, the statistic used is the cumulative monthly injury rate, which is expressed as injuries per million man-hours of exposure. This figure is obtained by dividing the actual number of injuries of a particular classification during a given month by the number of man-hours exposure for that month. (Total reported injuries are used in Exhibit 4.) These monthly rates are accumulated so that the January cumulative rate is the January

Exhibit 4
Sample Control Chart

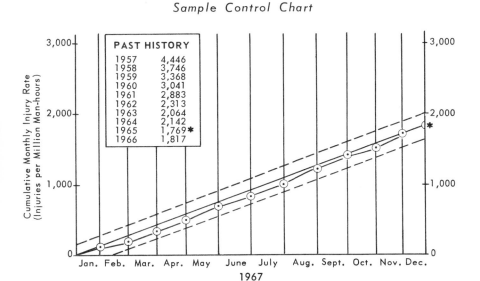

Exhibit 5
Seasonal Variation of Total Injuries Based on 15 years of data

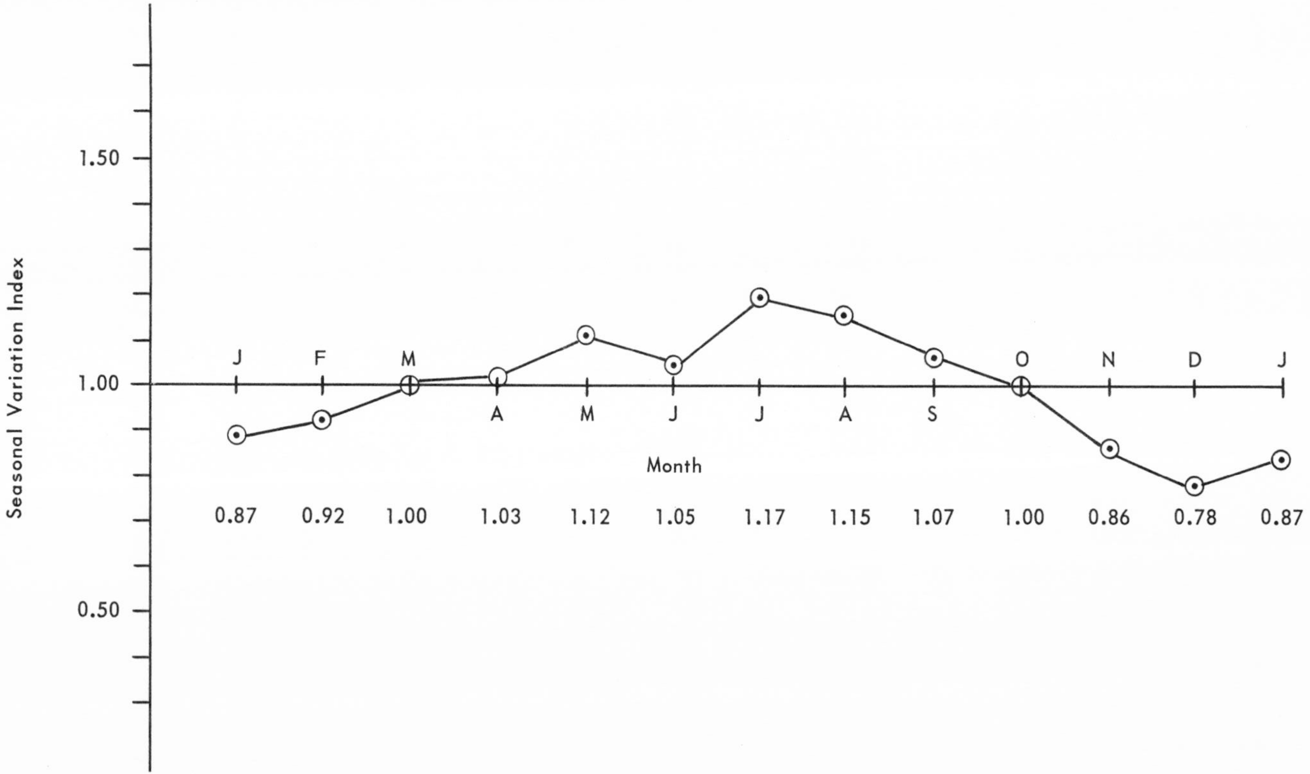

rate, the February cumulative rate is the sum of the January and February rates, and so on, until the December cumulative rate is the sum of the 12 monthly frequency rates. A simple plot of the total number of injuries would not show the rate. The average rate would result in diminishing control limits, thus covering up the changes. The use of the cumulative statistic permits immediate interpretation of significance of change.

Second, seasonal variation is removed to yield a linear trend. This is done by adjusting the spacing of the "month" coordinate of the chart. The width of the interval between the months is then a direct indication of the injury expectancy for that month. This spacing is based on the monthly averages for all previous years of history which are considered representative. For example, where five years of monthly frequencies are available, the average of all the Januarys divided by the average of all the months (60 in this case) represents the seasonal frequency index for January. The seasonal frequency index can be similarly calculated for each of the months and varied around a value of 1. A representative plot of the seasonal frequency index for total injury classification is shown in Exhibit 5. The index numbers for each month are noted. The individual monthly index is significant because it represents the relative probability of the occurrence of the particular classification of injury during that month and is, therefore, useful for predicting injury expectancies for each month.

Third, a goal can be set, and the progress toward meeting or surpassing the goal is easily understood at a glance. This encourages improvement and makes the need for corrective action visibly apparent. The goal can be the best previous year, last year's performance, or a percentage improvement of these. Whatever goal is selected, this control chart provides the guide for improvement.

The Standard Deviation

The control chart bears a set of confidence limits equally spaced on either side of the trend line to the goal for the year. These confidence limits are developed from the comparison of the observed injury frequency data with the average of that data. This comparison is called the "standard deviation" and is calculated by the following commonly accepted formula.

$$SD = \sqrt{\frac{\Sigma(X - X_{ave})^2}{N - 1}}$$

where SD = standard deviation.
Σ = summation of the quantities $(X - X_{ave})^2$.
X = observed frequencies.
X_{ave} = average frequency using all the observed frequencies or $\Sigma X/N$.
N = number of observations.

The standard deviation applies both above (more than) and below (less than) the trend line. This forms two lines parallel and equidistant from the trend line. One standard deviation (1 times the SD) gives confidence limits of 68 percent. That is, observed frequencies falling within the confidence lines are due to random distribution, and those frequencies falling outside these limits have odds of roughly 2 to 1 or better of being due to an actual change in the underlying situation causing the injury as opposed to normal chance variation around the trend line. When two standard deviations (2 times the SD) are used as shown, the confidence limits are 95 percent, which means that observed injury frequencies falling outside these confidence lines have odds of 20 to 1 or better of being due to some cause other than chance variation.

If three standard deviations (3 times the SD) are used, the odds rise to 100 to 1 or 99 percent confidence limits. These limits are very useful in determining whether apparent changes in the slope of the trend line are significant or not.

In Exhibit 4, these control limits with a trend line based on the best previous year or a percentage improvement over the best previous year are mathematically incorrect because the control limits are calculated on the average for a particular time period. To be technically correct, the control limits can be used only with a trend line that is based on the average during the same period. However, the error is small, and the use of control charts challenges the technical supervisor to seek improvement in this important phase of his responsibility.

Control charts are but pictures of performance; they require interpretation to obtain the maximum understanding and usefulness.

When interpreting control charts constructed on small sample sizes—such as 12 per year—there is a tendency to make premature assumptions. This can lead to decisions or program emphases that are not justified. This type of control chart provides two very important types of information for the accident prevention effort. The seasonal spacing foretells those months of the year during which a higher (or lower) frequency of injuries than average would be expected. This permits projects to be undertaken prior to these high frequency periods to reduce these frequencies. Such control efforts can be successful in reducing accidents of a particular type or in a particular work area or group during the period of emphasis.

The control chart also shows significant departures from the expected trend.

Examples of Seasonal Fluctuation

The control chart for total injuries (Exhibit 4) includes 15 years of data from a large chemical plant. Because of the volume of data included, the seasonal effects are not particularly pronounced. Control charts for major types or causes of accidents show significant seasonal effects. A chart of "slips and falls" (Exhibit 6) for this same plant shows that the highest probability for slip-and-fall injuries occurs in February and March and the least probability occurs in July. The supervisor, in planning his safety meeting program for the year, will stress this type of accident during the first quarter. He will use this cause as the subject of posters, promotional aids, and safety reminders in his area. He becomes keenly aware of this cause of accidents when he instructs his workers on the proper methods for performing work. During the summer months, he does not spin his wheels on these accidents during the low probability period. He efficiently applies emphasis on whatever needs attention at the moment.

This particular chart shows other interesting points about performance. The year's actual performance was not very good, as evidenced by the fact that performance lies above the trend line and outside the confidence limits. But it should be pointed out that this chart illustrates *trends;* thus the first and last quarters were the periods of poorer performance, whereas the second and third quarters were periods in which the performance was as good as the goal set, even though the plotted points are outside the confidence limits. In the proper interpretation of these points, this is very important. A casual glance suggests a bad year, but after a poor first quarter, control was reestablished, only to be lost again in the last quarter. And the supervisor, recognizing a problem of performance in the last quarter and facing a high-probability first quarter in the new year, renews and strengthens his effort on this accident category and its causes. To ignore this information can only result in continued poor, perhaps worsening, performance in a category that has the potential for serious and disabling injuries.

Notice the box at the upper left corner of Exhibit 6. Here the past history of this kind of accident is shown. The statistic used is the total cumulative monthly injury rate; therefore, the figures shown must be divided by 12 to obtain the average injury frequency.

The chart of injuries resulting from "temperature extremes" (Exhibit 7) is another example for interpretative discussion. This

Exhibit 6
Slips and Falls

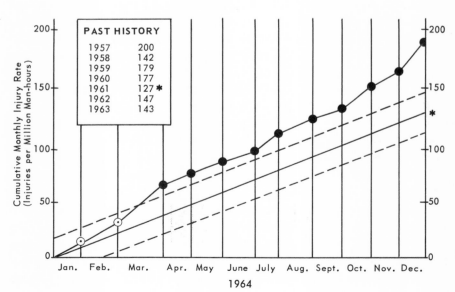

Exhibit 7
Temperature Extremes

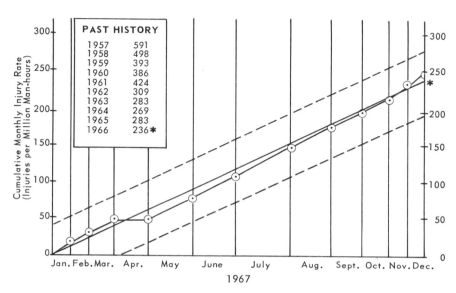

type of accident predominates in the summer months. In the early work on this program when the causes of these accidents were studied, it was learned that in the summer months a rash of burns to the hands and arms were caused by contact with hot lines. The workers wore short-sleeved shirts during the summer, and, although the injuries were not serious, a program was initiated to educate the workers about the hazards of hot surfaces. At the same time, exposed hot piping and other equipment was insulated. This proved more effective than asking these employees to be uncomfortable in long sleeves during the hot summer months. The frequencies of these injuries decreased, although the seasonal effect was not greatly changed. The reason is really simple—insulation of hot surfaces and recognition of the hazards reduced the exposure to this type of accident, not just in the summer but all year round.

The chart of injuries from "lifting, pushing, and pulling" (Exhibit 8) is another example. In past years, a very high probability for this type of accident was experienced in February. Supervisors spotted this, and they used the January meeting to stress this type of accident. The pressure was really applied, especially to craftsmen. The probabilities were upset, and the seasonal effect changed to a more random distribution. But note the abrupt worsening trend during the fourth quarter—a warning of trouble ahead.

When a particular cause was selected for program emphasis because of predicted high seasonal probabilities, the plant was very successful in reducing accidents during the period of emphasis. When that emphasis was dropped or shifted to some other cause, performance promptly returned to the predicted level. Are the control measures retained only during the period of emphasis or pressure? It would appear that this is a significant point.

The statistical information provided to supervisors may include tabulations of the types of accidents by work groups and by location. This gives the supervisor a complete breakdown of accidents resulting in injuries to his people and in his area.

This presentation of injury information is an effective tool in accident prevention. Managers and supervisors have a current breakdown of the injuries, a continuous graph of performance toward a goal, and, most important, a means to develop a funda-

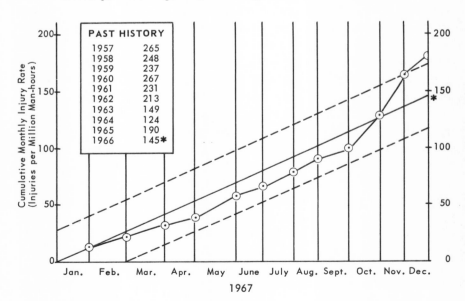

Exhibit 8
Lifting, Pushing, and Pulling as Accident Causes

PAST HISTORY	
1957	265
1958	248
1959	237
1960	267
1961	231
1962	213
1963	149
1964	124
1965	190
1966	145*

mental approach to future programs, action, and emphasis. The entire safety education program can be planned in advance with emphasis on the topics needing attention. Supervisors need no longer ask, "What should our safety meeting be about this month?" or "What posters should we use?" The subject is obvious and the results satisfying because now the supervisor knows the problem and can attack it with confidence. Managers and safety engineers can pinpoint particular trouble areas and bolster the weak spots.

Therefore, a plantwide problem can be recognized and solved. After this program-emphasis plan has been adopted, plantwide safety program periods—periods of extra plantwide effort—are observed. The film, "Knowing's Not Enough"[1] was shown in April 1959 to all employees in one plant, and the theme was emphasized heavily for the next three months. The plant had its best summer in years. When emphasis ceased, the pattern quickly returned to normal.

In 1960 a concentrated program was conducted to overcome the January effect, traditionally the poorest month in this plant's safety performance. For 31 days, every emphasis, gimmick, reminder that anyone could think of was used, and that particular January was the best since 1954—then a miserable first two weeks in February resulted until the normal predicted effects took over. Perhaps one lesson of this experience is not to stop safety campaigns, just let them fade away when they are replaced with others.

Performance Charts for Work Groups

The same type of performance chart can be used just as effectively for work groups—craft groups, laboratory groups, or any segment of the employee population engaged in a similar work function. The only limit to such use of the chart is that the groups must be large enough to generate meaningful and significant data. It is difficult to specify the exact minimum group exposure that will be statistically significant, because the frequency level is also a factor. Performance of groups has been effectively charted

---

[1] "Knowing's Not Enough" is a safety film produced by United States Steel Corporation and available through the PMA Corporation, 345 4th Avenue, Pittsburgh, Pa.

and controlled when the group total injury frequency is at least 100 (one injury per 10,000 man-hour exposure).

In small groups or in groups having a frequency of less than 100, we are dealing with isolated events separated by relatively lengthy time periods. M. J. Moroney uses the Poisson distribution as the statistical approach. In this approach we need only to know the average number of occurrences and to satisfy the condition that the expected number shall be constant from trial to trial in order to predict the probabilities of the occurrences. The actual frequency can then be compared with the predicted frequency to determine whether the trend is better or worse than expected.

The advantage of working with groups within a large plant lies in the ability of specific supervisors to become more effective in preventing accidents to employees directly under their supervision. Each supervisor may have a very different problem in his work group from that being experienced in other work groups or in the plant in general. The control charts he develops provide him with a tool for overcoming a specific hazard and an opportunity to show his concern for the safety and welfare of his people. Also, the manager can evaluate the performance of his supervisors in the important area of safety.

The use of these data as the basis for supervisory performance evaluation has certain pitfalls worthy of mention. The manager cannot honestly and fairly compare supervisors using these data alone, because work groups are different, work hazards are different, and individual supervisors are different. It is possible to evaluate the supervisor's concern for his people, his problems, and his responsibilities. If the manager uses the control chart to show that he would simply prefer fewer injuries without removing causes of accidents that may result in injuries; the supervisor will respond by reporting fewer injuries, but causes will go unrecognized and unsolved, and the result will be more injuries. The manager who understands control charts uses them to measure past performance to define problems and motivate action to prevent future accidents.

"Rolling" Control Charts

The calendar year is just a period of 12 months in the accident prevention business. If the charts are prepared on a calendar basis, the supervisor and manager must wait until mid-year to see what is happening. Also, there is little continuity from year to year. It is a little more trouble but certainly worth the effort to provide rolling control charts. These are prepared by adding a month and dropping a month, so that 12 months of performance are always available for study. Examples of such charts are shown in Exhibits 9 and 10. The seasonal variation is still the experience of past years, and the trend or goal line is simply the best previous calendar year or any other 12-month standard desired. The interpretation technique is the same as for calendar year plots, but the advantage lies in the availability of the immediate past 12 months for study.

The two examples shown were selected from a series of charts for various work groups in a chemical plant. Exhibit 9 is the control chart for a group of about 80 maintenance craftsmen. The accident expectancy is slightly higher during the summer months but is fairly uniform throughout the year. The group's performance is on goal and is much better than the previous year. The probability of an accident to a member of this group is about 50 percent of what it was eight years ago.

Exhibit 10 is the control chart for a relatively small group of operating and maintenance workers in an operating department.

Exhibit 9
Rolling Control Chart; Total Injuries for Boilermakers

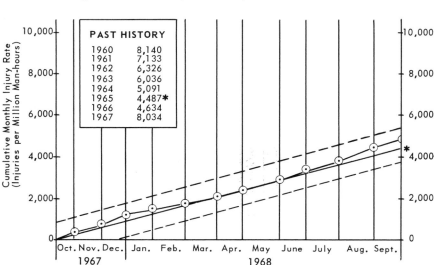

The wider spacing of the control limits is characteristic of a smaller group. The accident expectancy is pronounced in the months of March, April, and May. The group's performance for the years of 1966 and 1967 had definitely deteriorated, but control was reestablished in 1968—in fact, for the first nine months of 1968, performance was as good as during the best previous year (1965). Remember, it is the slope of the performance line that counts.

Perhaps the most common criticism of these control charts is leveled against the statistic of total injury frequency—the data base. This statistic includes all the very minor injuries, which may not have much potential, as well as the disabling injuries, which, by chance, have a high severity. It is a statistic that is subject to reporting variances, depending on the attitudes of supervisors toward reporting and the methods used in investigation. Another factor is the attitude of the employee who decides whether the injury is serious enough to report or whether it is an opportunity to quit work, get a smoke, or see the pretty nurse. He may even decide not to report if the first aid station is too far away or if it's raining. For these and other reasons, the percentage of actual injuries reported varies and challenges the validity of the data.

Exhibit 10
Rolling Control Chart; Total Injuries for Department C

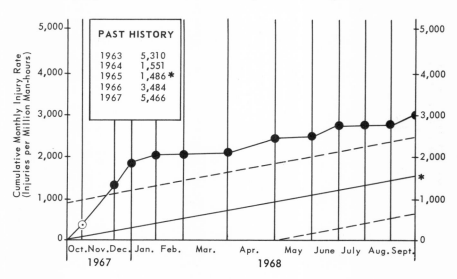

Accident Probability

To test the validity of these data, it is necessary to develop the concept of accident probability. Russell DeReamer stated that "the use of total injury frequency rates . . . provides a sensitive barometer of accident-prevention performances by sections of the plant or by supervisors." [2] All types of injuries have the same basic cause and differ only in severity. When an accident occurs, the severity of the injury result depends greatly on chance. Suppose a 24-inch pipe wrench falls from an upper level to the ground below. If no one is under the falling wrench, no one is injured, and we say a near-miss has occurred if we learn about the incident at all. But should a worker be standing under the falling wrench, a minor injury is sustained if the wrench grazed his hand; a serious injury is sustained if he luckily got off with a fractured finger; a disabling injury is sustained if the impact of the wrench broke his shoulder; or he may be killed.

William W. Allison gave a similar example when he wrote about an accident in which "a man sitting in the back of a truck is thrown out when the truck hits a bump. It is a matter of chance whether this unexpected incident results in a fatal brain concussion, a broken neck, a permanent disability, a minor injury, or (more frequently) no injury at all." [3] In each case, the accident and the cause are identical, while the result (the injury) differs in severity. If the accident has the potential to kill a man, it deserves the same investigative effort as if it had actually resulted in a fatality. It therefore follows that by reducing the causes of minor injuries, we are, in effect, removing the causes and reducing the probabilities of serious injuries, disabling injuries, and fatal injuries. If this is true, there must be a correlation or a probability relationship among these severity types of injuries. Studies have proved that such a relationship does exist. As expected, the probability decreases as the severity increases. Exhibit 11 shows the probability relationship among the severity types of injuries and the monthly seasonal probability variations for each severity type. This relationship over

---

[2] Russell DeReamer, *Modern Safety Practices* (New York: John Wiley & Sons, Inc., 1958), p. 303.
[3] William W. Allison, "High Potential Accident Analysis," *ASSE Journal*, July 1965, Vol. X, No. 7, p. 9.

Exhibit 11

Seasonal Variation of Injury Probability of Various Levels of Severity

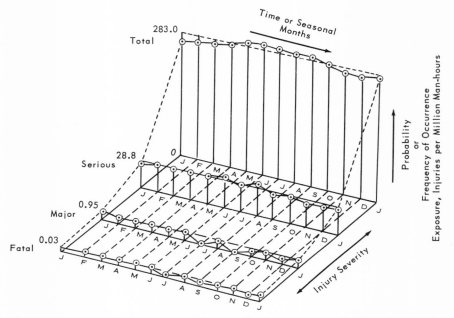

the past 13 years in a major chemical plant was such that for each fatality, there were 29 major or disabling injuries, 877 serious injuries (as defined by the Sohio Serious Injury Index Method [4]), and 8,618 total injuries.

Each industrial establishment has its own probability relationship. Differences in numerical value are the result of different types of work performed, different exposures to different hazards, different mixes of people, and many other differences among plants, industries, and locations. But the probability relationship will exist, and it is on this probability of injury occurrence that the preventive effort can be based.

The minor injury classification data may be adversely affected by two groups of employees—those who would not report a minor injury because of their disdain for the red tape of reporting and subsequent investigation, and those who would report anything simply as a means of getting away from the job whether the injury was job connected or not. These data are suspect if for no other

---

[4] A. R. Klingel, and O. C. Haier, "SOHIO Serious Injury Index," *National Safety News*, November 1956.

reason than it is impossible to obtain reports on every incident that results in injury. And, therefore, it is impossible to determine the percentage of minor injuries that is reported.

On the other hand, the serious injury classification offers the best data base because such injury cases are serious enough for the injured to demand attention, thus providing a more complete record of incidents. In most industrial establishments, there are a sufficient number of serious injuries to be significant for statistical purposes.

Serious injury frequency (SIF) has an extremely high correlation with total injury frequency (TIF). This is interpreted not as a cause-and-effect situation but as two effects sharing a common cause. This relationship for a large chemical plant is shown in Exhibit 12. This chart shows yearly data, which can be used to mark

Exhibit 12
Total to Serious Injury Ratio

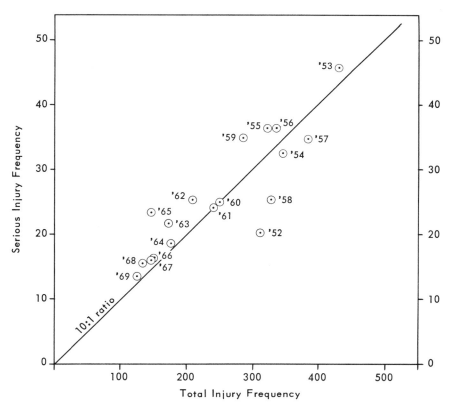

progress from year to year. Using all the monthly data for these same years, the TIF, SIF, and the ratio of these frequencies become sensitive indicators of current performance giving warning of trouble ahead. The prediction of a serious injury from TIF data or total injuries from SIF data carried a confidence of better than 99.9 percent, or only one chance in a thousand that there was no relationship between the two. In a similar manner, the relationship of major or disabling injuries to the SIF also exists, and there is less than one chance in a hundred of being wrong in saying that the serious injury and the major injury are related.

This relationship between these severity types or classifications of injuries changes slightly from year to year and more radically from month to month. Further, not all of these severity classifications vary in the same manner. Seasonal changes in probability or variations of probabilities with time do occur. These variations are important indicators, as will be discussed later, but the relationship between the severity classifications is valid, and through the control of the causes of accidents resulting in either the serious or minor injury, the causes of major and fatal injuries will also be controlled. And since these relationships do exist, the use of TIF as the data base for control charts is valid.

Control Charts as Trend Charts

Control charts, to be correctly analyzed, must be recognized for what they are, not what many people want them to be—a panacea for all problems. They are trend charts, indicating changes which must be confirmed by other inputs—data, impressions, opinion, observations, study, inspection, or whatever is available to determine whether a problem does or does not exist. These TIF charts show trends, therefore it is advantageous to have as much history to study as possible. This is why rolling 12-month charts are suggested. The manager has a year of data to observe, a seasonal index by months, and a comparison with past years.

Several factors about the plot must be fully understood before considered judgments or conclusions can be drawn. The slope of the performance line is the key. According to the method of plotting used, it is possible for the actual data to be well above and outside the control limits while the current performance is good (see Exhibit 10). This is true if nine to twelve months ago, a rash of injuries pushed the plotted points outside the limits, while in recent months (with control reestablished) the slope of the plot is equal to or less than the trend line. The analyst does not stop at this point, satisfied that TIF shows him in control. He keeps tab on the less frequent serious injury occurrence. He knows what his SIF has been over the years and that he is not experiencing an unusual frequency of serious injuries. Further, by looking at the seasonal spacings, he can recognize that the next several months are, by historical record, months of high injury expectancy. He knows that last year at this time, the group sustained a lot of injuries, more than expected, and was performing poorly. So the analyst recommends action to hold the trend in order that improvement can be realized.

In a group with a similar control chart but a current SIF significantly higher than experienced in previous years, the apparent good performance shown by TIF may be deceptive, and the causes of the various incidents should be reexamined to seek definition of the problem. On the surface, it would appear to be poorer performance accompanied by a lower level of reporting of minor injuries, but it could be a random rash of serious injuries with varied assignable causes that present no definite problem. A check of the attitude of supervisors and employees, an inspection of the area, and a communication session for the exchange of ideas (not necessarily limited to safety) would seem to be in order. If nothing seems to confirm the apparent performance, the supervisor has an opportunity to test his control. Faced with a couple of months of seasonal high accident expectancy, he communicates safety to his people through inspections to show his standards, he holds meetings to emphasize his concern and share ideas, and he confers with subordinate supervision to be sure that safety methods and practices are as required and expected. He tightens the performance standards without threatening the data and watches for the results. If he has been effective in raising the standard of performance, the results will be improved performance.

In this sensitive area of his responsibility, in which emotion over an apparent decay of performance may cause action to be ordered before the problem is defined, it is important that both supervisor and manager understand the limitation of control charts. Control charts neither solve nor define problems; they only show trends away from some preselected level of performance. Control charts tell the supervisor that he may have a problem. They are the tools by which change of performance can be detected so that control effort can be initiated effectively and efficiently.

TIF and SIF

One of the more interesting concepts of injury-based data for measurement is the relationship of TIF to SIF. Although a concept has developed that these two effects share a common cause and, therefore, a probability relationship, the use of the ratio TIF/SIF as an indicator of performance has not been discussed. We have said that each business establishment or plant has such a probability relationship and that the ratio is relatively fixed. One plant might have a 6 to 1 ratio while another has a 25 to 1 ratio, but, regardless of the numerical value, the ratio will be a fixed figure depending on the type of operation, the mix of people, and the uniformity of classification of injury severity.

If the operation is a machine shop or a steel fabrication shop, the ratio will be low. On the other hand, if the business is primarily an office function, the ratio will be high. In a more integrated operation consisting of both sizable office staffs and shop or assembly-line functions, the ratio will reflect the mix of functions and personnel. The uniformity of the classification of injuries as to severity is more important than the interpretation of some particular code of severity classification. As has been stated before, the comparison of current performance against previous performance of the same group is the test of the measurement system, not comparison with some other group, plant, or company. So it is not important whether a 12 to 1 ratio is good or bad compared with the plant across the street or some out-of-state company, even though they may be in the same business. Only the changes of ratio found in the same group using a consistent severity classification method are important.

For example, let us compare two plants and how they handle medical cases. One plant has a resident doctor who sees and treats literally every injury case. The other plant has a resident nurse, but the plant doctor is located several miles away in a neighboring community. When the injured appears at the first plant's medical station, the resident doctor will treat the injured. Thus, the literal classification is a serious injury because it was treated by a doctor. The resident nurse may treat the injured in the same manner, but the injury will be classified as a minor injury because the injury was not treated by a doctor. However, as long as the classification is consistently done in each plant, the value of the ratio is not lost.

Performance within a work group or plant will change, but the deviation may or may not be indicated in a TIF/SIF ratio change. That is, the ratio change may be a measurement indicator, and, again, it may not be. There are two variables that affect a ratio change—reporting level and performance. The reporting level, or the percentage of minor injuries actually reported, is difficult to determine and easy to change. It depends on the attitude of the supervisor—if he sincerely desires reporting, he will have a high reporting level in his group; but if he makes every incident a "federal case" or makes such comments as "stupid, careless, and foolish," he will have a low minor injury reporting level. When the boss is replaced, the employees will set a new reporting level appropriate to their appraisal of the new man. It is possible to confuse a performance change with a reporting level change, and the safety engineer must recognize and be alert to these effects.

Interpreting the Ratio

But if the reporting level is reasonably constant, then performance will show up in a changing ratio. When the ratio decreases, the severity of the reported injuries is increasing, and the probability of very serious disabling injuries is rising. Although this method is more qualitative than the more precise control chart method, it is an indicator worthy of attention and emphasis. When the ratio increases, the severity of the reported injuries is decreasing; generally, this is a sign of improved performance. However, increases greatly above the normal ratio should cause concern, because this change is more likely to be caused by a change of reporting level than by performance. Of course, monthly swings are quite wide, and small groups vary more than large groups.

It is imperative that when interpreting this ratio, the variations of TIF and SIF be considered. In studies of a plant with a 10 to 1 ratio experience, ratios of 9 to 1 to 12 to 1 did not appear significant, and even ratios outside these limits indicated a need to define problems or causes of the change only when they persisted for several months. At best, the ratio is an indicator, a qualitative device requiring other information to confirm the situation. By using the expected level of injuries, corrected for whatever seasonal effects there may be, the interpreter can draw conclusions from the actual data. If both SIF and TIF are rising, and the ratio of TIF to SIF is falling, watch out for trouble ahead. Usually, at this point, the manager wants to apply the brakes, so he lets it be known that the accidents have got to stop. He wants accidents to stop happening, but what he usually gets is a sharp reduction in reporting. Now the SIF is rising further because nothing has been done to get at the causes, and TIF is dropping because—well, because if the boss doesn't want accidents, the worker sure isn't going to report them. The performance worsens in an accelerated manner and positive managerial action must be taken to regain control.

On the other hand, a falling SIF and TIF and a stable or rising ratio are good news.

An analysis chart (Exhibit 13), which may be used for data interpretation, is provided with a word of caution: Those who manage data manage to get into trouble.

The point is that this workable method of measuring current performance against past experience provides a guide to the necessity for and degree of effort required to keep performance improving. But this method does not define or solve problems—it is intended solely as a measurement method. It is logically developed, statistically valid, and a valuable tool for the safety engineer who is reponsible for safety performance measurement and for the supervisor who is responsible for safety performance.

Reporting to the Manager

Communication of these data to the responsible supervisor or manager is vital to the accident prevention effort. If the safety engineer is not able to get the attention of the responsible supervisor,

Exhibit 13
Analysis Chart

SIF	TIF	Ratio	Performance
Down	Down	Up	
Down	Same	Up	Better
Down	Up	Up	↑
Same	Up	Up	│
Same	Same	Same	○
Same	Down	Down	│
Up	Down	Down	↓
Up	Same	Down	Worse
Up	Up	Down	

this makes the whole effort akin to academic exercise. There are many ways to effectively communicate the measurement results. One method is to telephone the particular manager and tell him that he has a problem that he'd better get to work on. You will have his attention but probably not his appreciation—in fact, he probably won't believe your assessment of his plight.

A more human and businesslike approach is recommended. Start with the assumption that the manager expects the staff specialist to assist him to do his job better, not tell him how to do his job. It would be better to provide the manager with a report on his performance backed by clear presentations of data to confirm that performance. A cover letter could include statements about the quality of his performance in relation to past performance and about the trend for the immediate future. The specialist should avoid comparing one manager's performance with another's—the purpose of performance measurement is to improve his performance, not to puff him up or deflate him.

There is always a reluctance to accept someone else's standard, and it is easy to find excuses for failure to be better than some other plant's performance. It is a frequent misuse of statistics to make comparisons of plants' or industries' safety performance with the disabling injury frequency. Aside from the fact that the plants and the industries are different in many ways, just consider the difference in interpreting the Z16.1 Code. Safety competition among work groups within a plant is unfair because of the many obvious differences in the groups. A plant or group always seeks improvement over yesterday's or even yesteryear's performance. To the supervisor who is responsible for 50 electricians, it is an interesting but unrelated fact that the welders had no injuries this month. He wants to know how his electricians performed relative to what they had been doing.

To the covering letter should be attached the control charts for the supervisor's group and a tabulation of the data for his area of responsibility. If the safety engineer makes himself available to discuss the charts and aid in the definition of problems, the manager will understand and appreciate this assistance. This approach can be applied to the plant manager and to each level of supervision in the organization. The report to the plant manager gives him a picture of the total plant—a statement of how things are going, with comments about apparent problem areas that need immediate attention and emphasis. It provides a series of control charts showing plant total and serious injury frequency, individual accident causes or types (thermal contact, slips and falls, striking against, struck by, and caught between, and so forth), and group performance (warehousemen, craftsmen, operators, laboratory personnel, and so forth). Finally, this report gives a tabulation of all injuries showing type, location, severity, or other breakdowns as are appropriate for the operation and organization. This will give the manager the total picture in a series of control charts from which he can adjudge performance at a glance.

To each lower organizational level, a similar report, or at least control charts, can be provided covering the performance of that particular group. For example, if a plant has ten functional sections, the report sent to the plant manager would include performance control charts for all the sections as well as one for the entire operation. The report sent to each of the section managers would include the plantwide performance chart and the section's particular chart. In this way, the plant manager knows what the plant picture is and can identify the pieces that make up that performance; the section manager knows what the plant performance is and what his contribution is to that performance.

Since there are many ways to present such information effectively, it would be presumptuous to recommend a "best" way. Staff specialists within an organization should know what methods are appropriate and successful for that particular business establishment. Control charts are an accepted and understood business tool. It need only be remembered that the success of control charts in performance improvement depends on how the charts are interpreted and used.

CHAPTER SEVEN

EXTENSIONS OF INJURY EFFECT MEASUREMENT

IN CHAPTER Six a measurement system based on incidents that resulted in injury was presented. In the struggle to develop this system, many tangents were followed. Some led to dead ends; some produced only interesting results; others led to unique uses of existing statistical methods in other disciplines.

The examination of "residuals" of monthly changes caused by seasonal effects showed apparent conflict between TIF and SIF. A residual is simply the difference between the predicted TIF (from SIF) and the observed or actual TIF. Since a probability relationship exists between TIF and SIF, there also exists a mathematical expression describing that relationship that might be written as

$$\text{TIF} = m(\text{SIF}) + k_1 + f(s)$$

where $f(s)$ is another expression describing seasonal or repetitive cyclical effects.

This is an oversimplification of a very complicated mathematical expression. Even so, the trend of performance is important, and significant deviations from this simple equation are of interest because they are the components of the trend and, as such, require an explanation.

Seasonal Effects

From the data given, it was found that TIF for the winter months was consistently lower than SIF would indicate, and, conversely, for the summer months, TIF was consistently higher than SIF would indicate. Perhaps during the months of inclement weather, the trip to the dispensary for a minor injury just wasn't worth the effort. Or perhaps the accident exposure of workers changes with the seasons, making the potential for serious injuries lower in summer than it is in winter. In at least one type of injury, that of strains (lifting, pushing, or pulling), one causal factor is that body muscles are tighter in cold weather. Lifting a 50-pound weight in winter might result in a serious injury, while the same act in warm weather might not result in any injury. The same is true for injuries resulting from slips and falls, which are more likely to occur during wet or icy periods. Conversely, thermal burns are more probable during warm weather but are usually less serious since most thermal burns are caused by bumping against a hot line or similar incidents that result in minor injuries.

Examination of 13 years of data showed that January is a bad month (having a probability almost three times higher than other months) for major (disabling) injuries. It is theorized that the letdown after the holidays, the after-Christmas bills, and the end of the annual safety program (for many companies) contribute heavily to laxness, to a lowering of the safety consciousness, and to a poorer attitude.

The effect of historical events was examined in some depth, and Exhibit 14 is the result of one such study. It is a chart of total injuries and exposure hours over a period of time. Two periods of note (1953 and 1957) occur immediately after a plant expansion and coincide with periods of recession and retrenchment in the business. The number of injuries were abnormally high. Does job security affect safety? Certainly safety consciousness, an attitude, is affected by the employee's environment and by changes affecting that environment. Such events as plant startups, major shutdowns, union contract negotiations, changes in plant policies or their enforcement, changes in plant management personnel, and promises of new units and job opportunities all apparently affect safety performance.

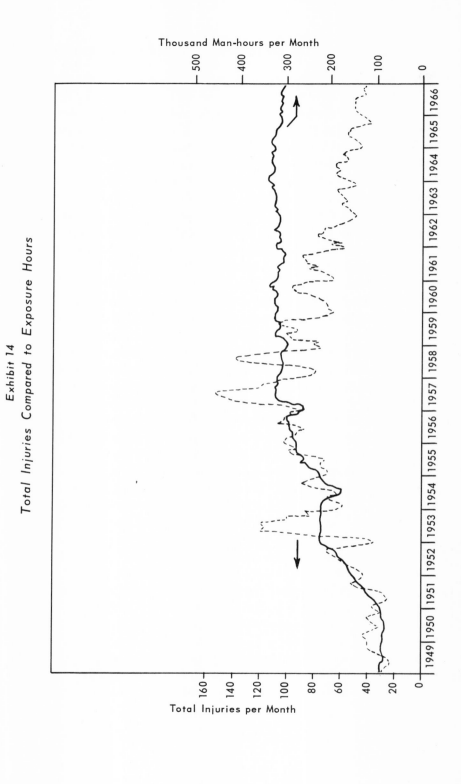

Gerard O. Griffin concludes, however, from a U.S. Bureau of Labor Statistics survey of manufacturing companies that the injury rate was higher in the summer months and lower in December and January and stated that "experienced safety directors know when employment and the mercury rise, so do work accidents." He also observes that "the injury rate often rises when there is an influx of new workers and experienced workers are shifted to new processes and activities." [1]

Ann Stresau from her survey conducted in the West Allis Works of Allis-Chalmers found that "rates of all kinds of (injury) cases and the severity rate are higher when employment is high and lower when the level of employment is low." [2]

P. J. Tuz and N. J. DeGrazia, in a comparison of a leading business cycle indicator (average workweek of manufacturing workers) and disabling injury frequency rates for the decade of 1955 to 1966, noted that "a close relationship exists between the economic indicator and the frequency rates in terms of similar variation" and that "the similarity is remarkable." They further observe that "the same similarity exists if we compare frequency rates to changes of employment or changes in overtime hours for all manufacturing." [3]

Exhibit 14 plots the results of Monsanto's Texas City plant's loss prevention effort. Injuries were proportional to exposure hours, except during the past seven years. During this period, there was a significant reduction of injuries (50 percent lower) while exposure remained almost constant. There were two reasons for this change: First, the control system described in Chapter 6 was installed, and, second, control systems were installed in other areas of managerial responsibility resulting in similar gains in cost and quality.

Once again, the reader should remember that it is control, not measurement, that gets results. Measurement is based on the results or effects of the incident; control must be based on the causes of these incidents.

John V. Grimaldi stated that results from a study of 15 businesses "broadly indicate that in a business where there is fairly good control of all contingency costs, a generally good safety profile will

---

[1] Gerard O. Griffin, "Little-Known Facts About Injury Occurrence," *National Safety News*, May 1960, p. 63.
[2] Ann Stresau, "Safety—Reliability," *ASSE Journal*, January 1966, p. 16.
[3] P. J. Tuz and N. J. DeGrazia, "The Socio-Economic Implications of Industrial Safety," *ASSE Journal*, September 1967, Vol. XII, No. 9, p. 23.

prevail."[4] He later stated, "it appears that the safety-efficiency couple is more properly described by saying that advancing managerial efficiency improves the accident prevention effort" and that "safety may be more a function of general managerial effectiveness in controlling operations than safety specialist activity to persuade employee safety awareness."[5]

Several statistical methods have proved most useful in understanding the content and meaning of the available injury data.

Cumulative Sums Techniques

Cumulative sums techniques have been used effectively in several ways. For instance, in groups for which the exposure is either too small for significant confidence limits or is not known, cumulative sums of the deviation from some base of the number of injuries are helpful in showing trends. In this case, the measurer may assume a base number either at random or at the past average of the group. He then obtains the positive or negative difference between the actual number of injuries and the base number and adds this difference number to the numerical value of the previous point to give a new value. An example of the calculation is given in Table 1.

Table 1
CUMULATIVE SUMS METHOD
(Base = 10 injuries per month)

Month	No. of Injuries	Difference from Base	Cumulative Sum
Jan.	12	+2	+2
Feb.	7	−3	−1
Mar.	9	−1	−2
Apr.	10	0	−2
May	15	+5	+3
June	16	+6	+9

[4] John V. Grimaldi, "Appraising Safety Effectiveness," *ASSE Journal,* November 1960, Vol. V, No. 4, pp. 57–62.
[5] "Another Look at Stimulating Safety Effectiveness," *ASSE Journal,* April 1962, Vol. VII, No. 4, pp. 21, 23.

In this type of data presentation, the analyzer must remember that the actual numerical value of the cumulative sum (right-hand column) has little or no significance except in relation to the preceding values, and this relationship (how many units and in which direction) depends on the base used, since each value is calculated from the difference of that month's data from the base. Also the analyzer must understand what data are being used (number of injuries is different from frequency of injuries), what base is used (random, previous year, previous quarter, best time period, or whatever), and, most important, that the only information available from a plot of these cumulative sums is the trend (which way it is going, how fast, and whether this direction is good or bad).

Cumulative sum plots are very useful because they are easy to construct and they show trends. It doesn't take a fancy, expensive computer to develop the data. Anyone who can add and subtract and perform simple plotting can build a meaningful presentation of data analysis. This method of data handling is particularly useful for small groups in which the number of injuries is not large, and frequency is low, or either of these factors vary widely month to month. They are useful for "quickie" plots of this year's trend as compared to last year's average. Such data as doctor cases, nonoccupational injuries, and other emergency events (fires, spills, utility failures) can be plotted to give quick, meaningful trend pictures.

The Weighted Severity Index

Another very useful, but less scientific, method for recognizing performance trends is that which we call, for lack of a better term, the Weighted Severity Index. One study of a large group of injury cases found that the severity of cases of certain accident types was significantly higher than for others. For instance, a higher percentage of the lifting, pushing, pulling cases are serious and compensable (under the State Workmen's Compensation Act) than of such accident types as temperature extreme, striking against, and others. It was reasoned that if the severity could be weighted so that the accident types with low severity potentials could be weighted low and those with high severity potentials weighted high, a Weighted Severity Index could show how the accident potential was acting.

In Table 2, this weighted severity factor is developed. This table

Table 2
WEIGHTED SEVERITY INDEX FACTORS

Accident Type	Injury Cases Percent of Total	Percent of Serious	Percent of Compensable	Total Weighted Severity	Weighted Severity Factor
Striking against	29	13	7	461	3
Struck by	26	31	13	890	6
Caught between	8	11	4	287	2
Slips and falls	6	10	15	771	5
Chemical contact	9	9	7	405	3
Temperature extreme	13	9	9	499	3
Lifting, pushing	8	15	44	2,123	14

is the result of five years of data from a plant with 2,000 employees. All of the injury cases were classified by type and the percentages calculated for the total cases, the serious cases, and the compensable cases.

Other accident types did not occur frequently enough to be statistically significant.

For the period studied, the severity ratio was as follows: For each compensable case, there were 5 serious injury cases and 45 total injury cases. The weighting was done by assigning the total injury case a value of 1, the serious injury case a value of 9, and the compensable injury case a value of 45. This is simply the application of a weighting factor on the basis of the occurrence of the injuries. Total weighted severity was then obtained by multiplying the percentages for each accident type by the appropriate weighting factor and adding these together to get 461 for striking against, and so forth. By simplifying these numbers to give a ratio (in whole numbers) among the accident types, the final column, the weighted severity factor, was obtained.

Each month, as the injury cases are classified by accident type, the total number of cases for a particular type can be multiplied by the factor for that type. By adding all these products (one for each accident type), a monthly severity number is obtained. If frequency is preferred, the severity number can simply be divided by the exposure in million man-hours. This severity number or frequency is the Weighted Severity Index. The index can be handled in any of the methods already described. For a large plant with computers available, the index can be calculated on serious injuries and control charts developed from the data, or, using total injuries, control charts for groups can be developed. Although it should not be inferred that this index is better than the Serious Injury Index, it has one advantage. The Weighted Severity Index can be based on the same serious injury information and can, in addition, give weight to those types of accidents that have the potential of severely hurting or disabling people. The index will show when the potential severity of injuries is rising in a more dramatic manner than the Serious Injury Index does.

For a small plant or one in which the necessary manpower and computers are not readily available, using a simple plot of the index based on total injuries or the cumulative sum method, trend curves can be easily drawn for presentation to management. This is one of the best measurement methods for the small plant or the part-time safety engineer because once the index factors are calculated, the repetitive monthly chart is easily drawn and makes a professional impression.

The Rolling Average Method

Another useful technique of data handling is the rolling average method. To begin with, a time period is selected (statisticians generally use 5 months, 12 months, and 15 months). For example, assume a time period of five months is applied to total injury frequency. Table 3 shows the calculation for a year.

Of course, if TIF is known for the last four months of the previous year, the gaps in Table 3 can be calculated. The average of the sums or the five-month rolling average is the number to be plotted. In this case, the performance is improving. Again, the

Table 3
ROLLING AVERAGE METHOD

Month	TIF	Sum of Current and Previous Four Months	Average of Sums
Jan.	150		
Feb.	220		
Mar.	175		
Apr.	185		
May	120	840	168
June	160	860	172
July	135	775	155
Aug.	120	720	144
Sept.	145	680	136
Oct.	130	690	138
Nov.	100	620	124
Dec.	140	635	127

method is easy to use, and, since it shows trend and smooths out the data without losing significance, it results in meaningful data presentation.

One of the practical uses of this method is shown in Exhibit 15. This exhibit plots the 12-month rolling average for both TIF and SIF for many years. Since a probability relationship exists between TIF and SIF, plotting the rolling average of each, using a coordinate that makes the two frequencies coincide (or fall on top of each other) according to that probability relationship, will show deviations from that relationship as well as overall trends. The periods indicated by arrows are examples of poorer performance because the severity of injury is increasing (higher SIF), while total injury frequency continues its downward trend. The scatter of monthly data has been smoothed by the rolling average for better readability of the trend of the data.

At the same time, a change is apparent (whether that change is in SIF, TIF, or both). This chart shows periods of performance change even though the overall trend is improvement. It warns of possible deterioration of performance so that corrective action can be initiated. Yet the rolling average is not so sensitive that it moti-

Exhibit 15
SIF-TIF Rolling Average (12 Months)

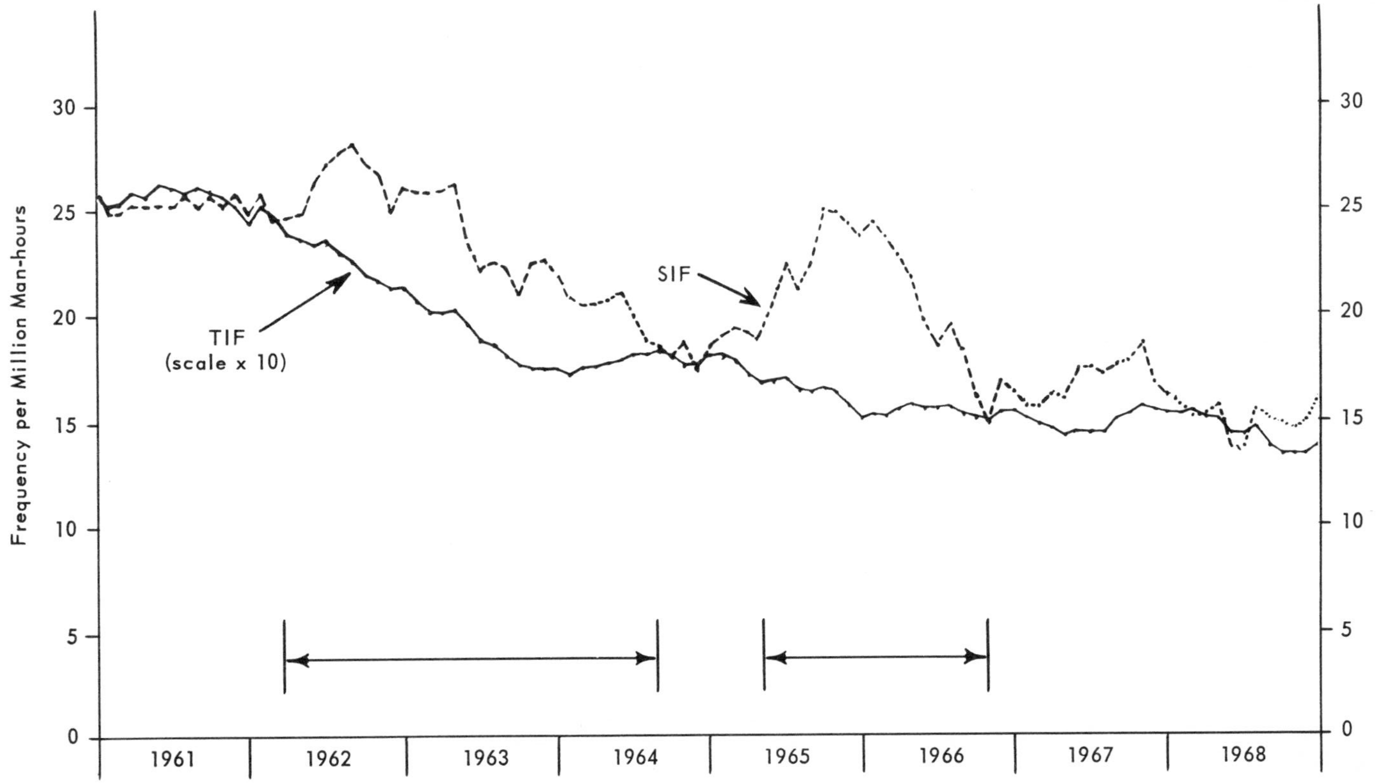

vates the sort of panic crash programs that are rarely justified. The safety engineer, who would like an ace up his sleeve, might use the twelve-month rolling average for presentation to management and develop the five-month rolling average for his own counsel. This would show management the trend and would give the safety engineer the sensitivity of a shorter-range rolling average.

Of course, more complicated techniques are available. Most of these are used in other disciplines, but not in safety performance measurement. Two such methods will be discussed because they *have* been used in safety performance measurement, with potential application in one case and success in the other.

The Seasonally Adjusted Economic Series

The first of the two is the Seasonally Adjusted Economic Series developed by Julius Shiskin for computing the seasonal, cyclical, and irregular component adjustments of economic indicators used in business analysis. This tremendous achievement in economic analysis not only has brought deserved honor to Shiskin but is widely used, in many modified series, by government and business. It is used in stock market analysis, consumer goods market analysis, banking, census and tax studies. It has been programmed into literally thousands of series and is well known to statisticians.

Monsanto's limited experience with a modification of this method found it to be useful. Seasonal effects were confirmed. Cyclical variations were recognized. The irregular component was found to vary at a constant level and represented a small component of the data. This method's value as a safety measurement tool remains uncertain because, with only one variable (TIF or SIF) processed, the method was too complicated to justify its use. However, the technique probably offers an opportunity for the cooperative efforts of statistician and safety engineer to develop the dependent variables and, subsequently, the projected trend of performance.

The Learning Curve

Another technique which has received considerable study by Monsanto is the method classically called the learning curve. This method has been used extensively in production cost analysis. It was first developed by Dr. T. P. Wright for cost analysis of airplane production. He observed that the cost of producing each of a series of orders for airplanes of a particular model diminished as the orders were filled.

Experimental application of the learning curve has shown the effect of learning on the repetitive assemblies of equipment, the effect of incentives on productivity, the effect of low- and high-volume production, and other productivity situations. A logical use of this method would be to determine the obsolescence of manufacturing processes or the point at which cost improvement ceases unless a significant change is made in the process or that process is replaced with a more modern, lower-cost process. Winfred B. Hirschmann comments on the fact that learning curve techniques, although known for a long time, have not been adopted widely. He discusses the use of this technique in petroleum processing, construction, and inflation and obsolescence control.

Safe performance of work is also a learning process. Experience should teach us to do a better job of sweeping the floor, operating a massive crane to lift and place steel, or driving a truck. But people don't always learn. The job becomes routine and boring, corners are cut, procedures are violated, and accidents occur. Good safety programs appear so effective that change seems unnecessary, but even the effective ones can lose their punch. During orientation, workers are informed of the safety rules, but these are easily forgotten. Safety performance should improve as experience increases, programs improve in quality, job procedures are refined, and effort is applied. It seems logical that if production costs decrease as we learn how to produce more efficiently, safety performance should improve as we learn how to perform work more safely. Neither just happens; it takes a concerted effort on a continuous basis to make it so. The learning curve is just a method to chart the progress of that effort.

Dr. Wright suggested that the mathematical model for the learning curve is

$$yi = ai^{-b} \tag{A}$$

where yi is the cost of the ith unit.
    a is the cost of the first unit; therefore $a = y$.
    i is the production count beginning with the first unit.
    b is the measure of the rate of reduction.

For the learning in safety performance, using SIF as the measurement unit, the equation becomes

$$\text{SIF}_t = \text{SIF}_0 t^{-b} \qquad (B)$$

where SIF_t is the frequency at time, t.
SIF_0 is the frequency at the beginning of time.
t is accumulated man-hours since beginning; it can also be shown as million man-hours or dated years on graph.
b is the measure of the rate of reduction.

As in the original learning equation (A), this model has the characteristic of describing constant percentage reductions. Each time (man-hour) increase of a constant percentage sees an accompanying injury frequency decrease of a constant percentage.

If t_2 and t_1 are two points in the exposure history of a work group, then

$$\frac{\text{SIF}_2}{\text{SIF}_1} = \frac{\text{SIF}_0 t_2^{-b}}{\text{SIF}_0 t_1^{-b}} \qquad (C)$$

$$\frac{\text{SIF}_2}{\text{SIF}_1} = \left(\frac{t_2}{t_1}\right)^{-b} \qquad (D)$$

The log transformation of the original safety equation (B) is:

$$\log \text{SIF} = \log \text{SIF}_0 - b\,(\log t)$$

which is the equation of a straight line with slope $-b$.

The log transformation of equation (D) is

$$\frac{(\log \text{SIF}_2 - \log \text{SIF}_1)}{(\log t_2 - \log t_1)} = -b = \text{slope} \qquad (E)$$

The learning curve can be constructed on log–log graph paper on which SIF is plotted on the vertical ordinate and the accumulated man-hours on the horizontal ordinate. If the improvement in performance was steady, SIF values plotted would form a straight line of negative slope. This slope is the rate of learning or rate of improvement. These rates are expressed as a percentage of the no-improvement level of 100 percent. In other words, a learning rate of 80 percent is an improvement of 20 percent and has a slope equivalent to a reduction of 20 percent from the initial value or a

Exhibit 16
The Learning Rate Curve

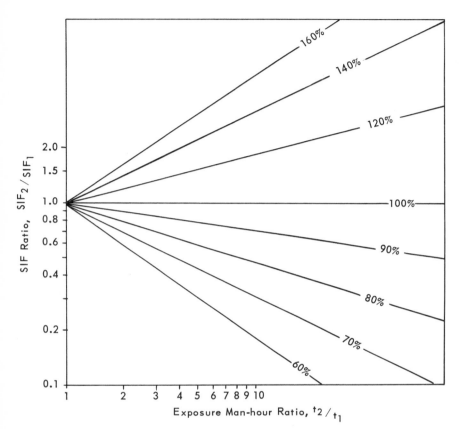

new value of 80 percent of the initial value. This is shown in Exhibit 16, which is an adaptation of the curve developed by Hirschmann.

The rate of progress of the learning is usually described as that reduction of injury frequency (or whatever) which occurs when the time quantity is doubled. This then is equal to a quantity C^{-b} expressed as a percentage, where $t_2/t_1 = C = 2$.

Then

$$C^{-b} = LR \qquad (F)$$

where C is t_2/t_1, the time ordinate ratio.
    $-b$ is the slope of the plotted curve.
    LR is the learning rate expressed as a decimal.

Making a log transformation and substituting the value of $C = 2$

$$-b(\log 2) = \log LR \quad (G)$$

and substituting (G) into (E), we obtain the equation

$$\log LR = \frac{(\log 2)(\log \text{SIF}_2 - \log \text{SIF}_1)}{(\log t_2 - \log t_1)} \quad (H)$$

This equation permits, by transposition, the calculation of reasonable expectations of injury frequency at some future time so that long- and short-range goals can be set and performance charted against those goals, as shown in Exhibit 17. The equation for predicting or setting a goal in the future is a transformation of (H):

$$\log \text{SIF}_2 = \frac{(\log LR)(\log t_2 - \log t_1)}{(\log 2)} + \log \text{SIF}_1 \quad (I)$$

Let us take, for example, a plant which accumulates two million man-hours each year. On January 1, 1968, the plant had accumu-

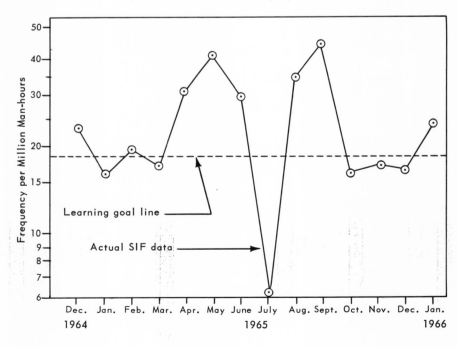

Exhibit 17
1965 Safety Performance
(Based on Serious Injury Data)

lated 8 million man-hours and had an average SIF of 30 serious injuries per million man-hours for the last quarter of 1967. (Point data for one month are quite variable, so a quarter is used in this example to give a truer picture of the performance at that time.) By January 1, 1969, 2 million man-hours more have been accumulated, bringing the total to 10 million man-hours since operation began. SIF for the last quarter of 1968 was 27. What has been the learning or performance rate?

Using equation (H)

$$\log LR = \frac{(\log 2)(\log SIF_2 - \log SIF_1)}{(\log t_2 - \log t_1)}$$

$$= \frac{(0.30103)(1.43136 - 1.47712)}{(1.00000 - 0.90309)}$$

$$= \frac{0.30103\,(0.04576)}{(0.09691)}$$

$$= 0.14214$$

$$LR = 0.721 \text{ or } 72\%$$

Assuming the goal for improvement was to continue through the coming year at the same rate, what is the goal for SIF at the end of 1969? Using the transformation (I) and assuming that two million man-hours will be worked during the coming year, we have

$$\log SIF_2 = \frac{\log LR(\log t_2 - t_1)}{\log 2} + \log SIF_1$$

$$= \frac{(-0.14214)(1.07918 - 1.00000)}{(0.30103)} + 1.43136$$

$$= 1.39397$$

$$SIF_2 = 24.8$$

The same result ($SIF_2 = 24.8$) would have been obtained if SIF_1 of 30 at t_1 of 8 million man-hours had been used.

Is this a good learning rate, or is improvement too slow? Any improvement is in the right direction. Studies generally agree that there is no universal curve that fits all learning, nor is there a learning rate that can categorically be ascribed to a particular learning situation. However, the learning curve provides the ability to recognize the existence or absence of progress and the desirability of measuring that progress. Improvement is important. When the learning curve flattens (100 percent) or goes up (greater than 100 percent), that plant or group has quit learning (improving) and corrective action is in order.

This learning curve method acknowledges one of the facts about safety performance that is often overlooked in setting goals. Performance, no matter how good, will never reach that unattainable but nonetheless desired level of zero accidents. It provides a method of setting a practical goal for improvement and a plot of progress toward that goal, as shown in Exhibit 17.

Sociologists say that learning at the 80 to 90 percent rate is quite admirable for a mix of men and machines. So a very reasonable goal for safety performance improvement would be 10 percent per year. A plot can be prepared to show this goal by using log–log graph paper and plotting SIF on the vertical ordinate and man-hours exposure along the horizontal. As time progresses, the actual values of SIF can be plotted to show their relation to the goal line. Due to the wide variations in monthly SIF values, the plotted points may not show any particular relationship to the goal line, or, as the statistician would say, there is a lot of "noise" in the data.

Multiple Linear Regression Techniques

So that the learning can be more meaningful, Multiple Linear Regression (MLR) techniques have been applied to the SIF data. This is simply a method of finding the best curve fit for the data—the line that best describes the trend of the data. Such techniques are well known to computer-oriented personnel and programs are available. For those without a computer, more tedious manual methods are available, but even an "eyeball" view of the plotted data will indicate a trend. Of course simply using a rolling three-month average can do wonders to smooth out the data variations.

Whether performed by hand or computer, the interpretation of the curve is the payoff. As with the control chart and cumulative sum techniques, it is the trend that must be read and a change is significant. The learning curve is sensitive at the current end of the data. There is little significance if a single point is up or down from the trend of the curve, but when readings of a three-month period are higher than the slope that has been observed, the observer should be concerned, and five months of higher readings should get him on his feet to go to work on problem definition and correction.

For those who react emotionally to a bad month, this method can only cause ulcers, but for those who understand that progress is made over the long run through fundamental programs for continued improvement, this method will prove a very useful tool for safety measurement.

For those who are fortunate in having computerized Multiple Linear Regression programs, this method, when applied to both SIF and TIF monthly, can give the safety engineer real insight on matters as they are and as they will be. Like any trend method, it does not predict but only *indicates* a trend, assuming nothing happens to change that trend. By using all available data, in this case monthly data for SIF since 1952, the learning curve can be produced each month. Exhibit 18 shows a plot for the years of 1952 through 1957. The residuals (the differences between the actual value and the MLR calculated value) are charted in the middle of the figure, and the cumulated residuals are plotted at the bottom. The chart and curve construction in Exhibit 18 is tedious to do without a computer, but this presentation is clearer than a computer printout for purposes of this discussion. This type of data presentation is appropriately entitled a "profile" because it shows the features of performance. The top curve—the MLR handling of SIF data using the learning curve model and plotted on log-log coordinates—shows an improving performance after a period of poor performance. It would appear that performance was poorest in 1954 and improved steadily thereafter. This is a problem of interpretation. Because the first year (1952) is not significant in these plots, the conclusions in this presentation begin with 1953. Performance was poor in 1953, and each chart shows this—a higher than 100 percent learning; positive residuals, quite pronounced in all the early months; and a fast-rising cumulative residual caused by the repetitive, significantly positive residuals. But in the latter part of 1953, performance began improving and did so for the remainder of this time period except for a modest upset in late 1955 and early 1956. Note the tendency of the learning curve to turn up (indicating poorer performance) at the end of 1957. This may or may not really mean that performance was poorer. The swings were quite wide in mid-1957, suggesting

Exhibit 18
A Monthly Serious Injury Profile

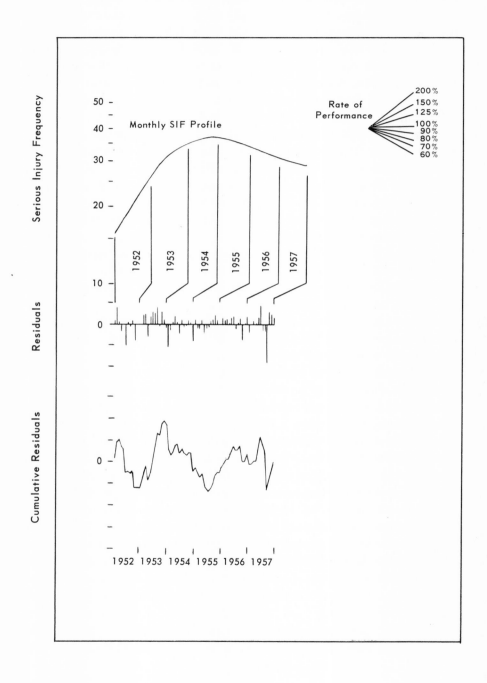

that some upsetting causes were active. The three positive residuals in the last quarter of 1957 are applying pressure to make the curve turn up.

Exhibit 19 is a one-year (1958) extension of Exhibit 18. The same shape of the learning curve shows that improvement began in late 1953. The upset from the end of 1957 through most of 1958 was offset by an improved last quarter in 1958, which gave the learning curve the appearance of smooth, continuing improvement. Actually, the curve computed at the end of the third quarter showed a marked swing upward confirmed by the upward movement of the cumulative residuals at that time.

Exhibit 20 extends the curve through 1959. The learning is significantly poorer. The improved performance of late 1958 was only temporary, and 1959 was a year of generally poorer performance as compared to earlier years.

Exhibit 21, in which 1960 has been added, shows considerable improvement. Although the curve still swings upward, it is an improvement over the end of 1959. And Exhibit 22 shows continued improvement through 1961.

Exhibit 23, in which 1962 has been added, appears to show continued improvement, but after two and a half years of improvement, the second quarter of 1962 was actually a nightmare. In January 1962, a disabling injury brought to an end the longest string of safe man-hours (6,814,000) in the history of this plant. It seemed as if all the serious incidents contained during the many months of improving performance were let loose to even up the score. There were several events that, in addition to the termination of the long safety record, probably led up to a plantwide letdown. A devastating hurricane, Carla, occurred in September 1961 and resulted in personal property loss to a majority of employees. A hard freeze, uncommon in this southern area, hit in January 1962 and caused another plant shutdown and long hours of work and strain on the employees. With these events behind them, the letdown was a logical result. After a concerted effort inspired by a timely recognition of the problem, the plant recovered and started the improvement course again. Without the upset in mid-1962, the curve would have been much better.

Exhibits 24 and 25, showing the addition of the next two years, indicated marked improvement. Exhibit 26, on which 1965 was

(*text continued on page 108*)

Exhibit 19
A Monthly Serious Injury Profile (Extended One Year)

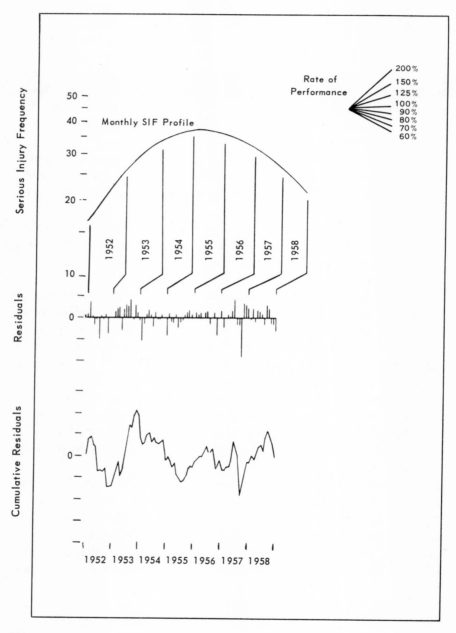

100

Exhibit 20

A Monthly Serious Injury Profile (Extended Two Years)

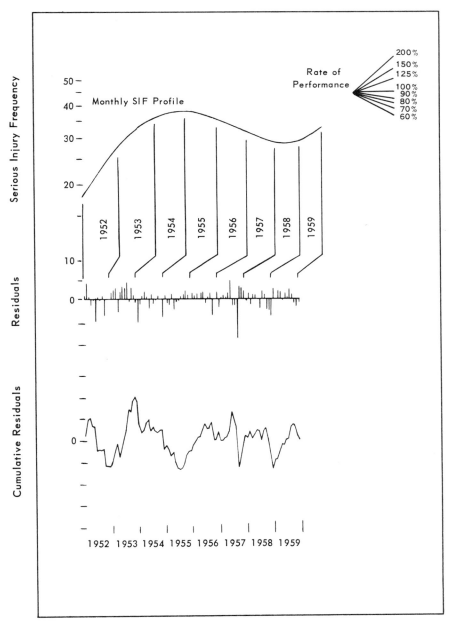

Exhibit 21

A Monthly Serious Injury Index (Extended Three Years)

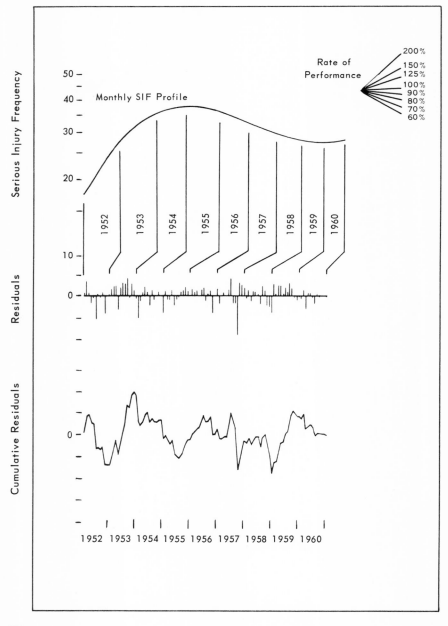

Exhibit 22

A Monthly Serious Injury Index (Extended Four Years)

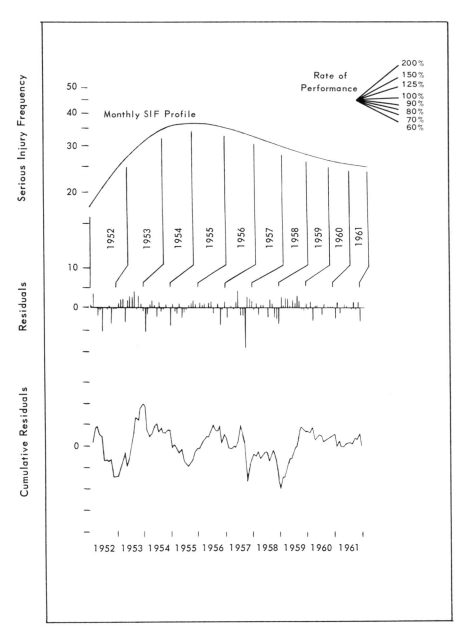

Exhibit 23

A Monthly Serious Injury Index (Extended Five Years)

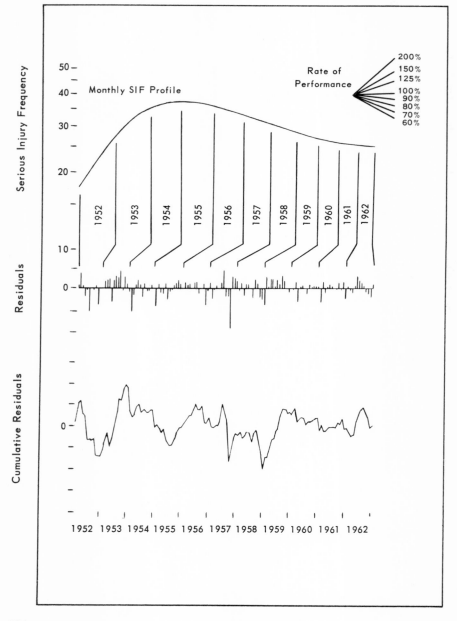

Exhibit 24

A Monthly Serious Injury Profile (Extended Six Years)

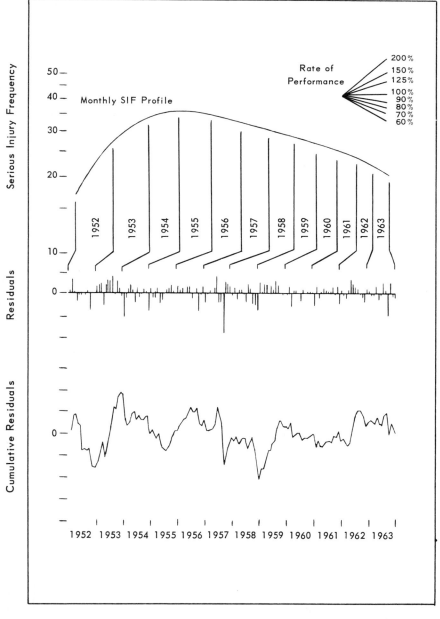

Exhibit 25

A Monthly Serious Injury Profile (Extended Seven Years)

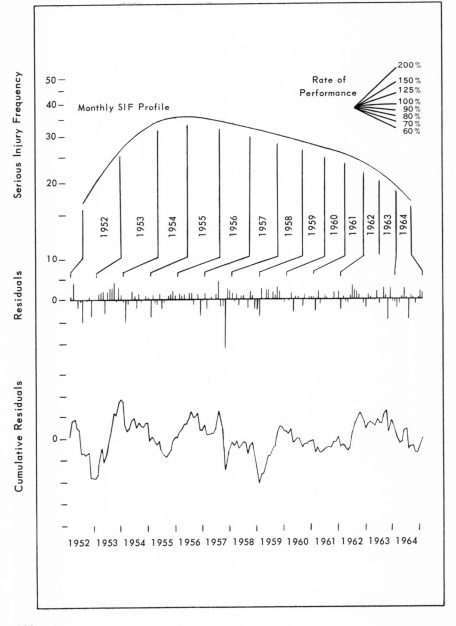

Exhibit 26

A Monthly Serious Injury Profile (Extended Eight Years)

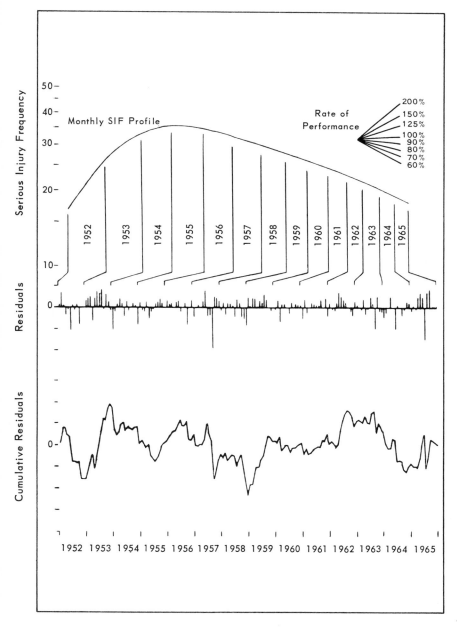

added, shows a reversion to poorer performance again. Note the positive residuals.

By adding the computer card containing the current month's SIF, a printout of the results of MLR can be obtained each month. Examination of the value of the last point of the learning curve, the slope of the last several points, and the direction of the residuals can give early indication of a change.

Again, a warning is in order. If the trend has been showing improvement and the current month has produced a very high injury frequency, the MLR will show an upward swing after all the data are computed. The analyst should wait for another month, because if the next month is back on the improvement trend, the MLR will recognize the poor month for what it really is—a temporary upset—and will give credit to the general improvement trend. The time to get excited is when, after three or four months, the tail of the MLR curve continues its change of slope in the less favorable flat or upward trend. Then it is time to look for reasons for the change and apply short-term emphasis to bring the performance back on the improvement trend. The work-group TIF control charts and the information gained from inspection and discussion will define the problem and, when corrective action is applied, improved performance.

Short-range Prediction

In the search for a technique of short-range prediction, the MLR seems to have promise. The learning curve method with the data handled by regression gave smooth curves and was very sensitive on the current end of the data; that is to say, a very high or a very low frequency in the last month would really "whip" the end of the curve. But even with this whip effect, the curve was smooth. However, if the MLR method were applied to just the most recent 12 months of data, the curve would be quite different from the curve for the same period where all the data had been used in the regression program. A study was made of 9 years of data in which the MLR method was used on the first 12 months of data and then repeated on the rolled 12 months of data; that is, a month was dropped and a month was added until data for 108 such curves were available. This was done for both SIF and TIF for this nine-year period. The following very interesting information was gained.

1. The general slope of the 108 plots of 12 months each was the same as had been obtained from the overall learning curve.
2. The 12 points for any given month obtained from the various MLR calculations were quite close in value, so that although there were 12 different curves passing through a given month, the curves tracked very well.
3. TIF and SIF followed each other as they increased and decreased throughout the period, again confirming that a probability relationship did exist.

Plots of two recent years of these data are included in Exhibit 27, which shows the composite curves for SIF, and Exhibit 28, which

Exhibit 27
SIF Trend Profile Using 12-Month Learning Curves

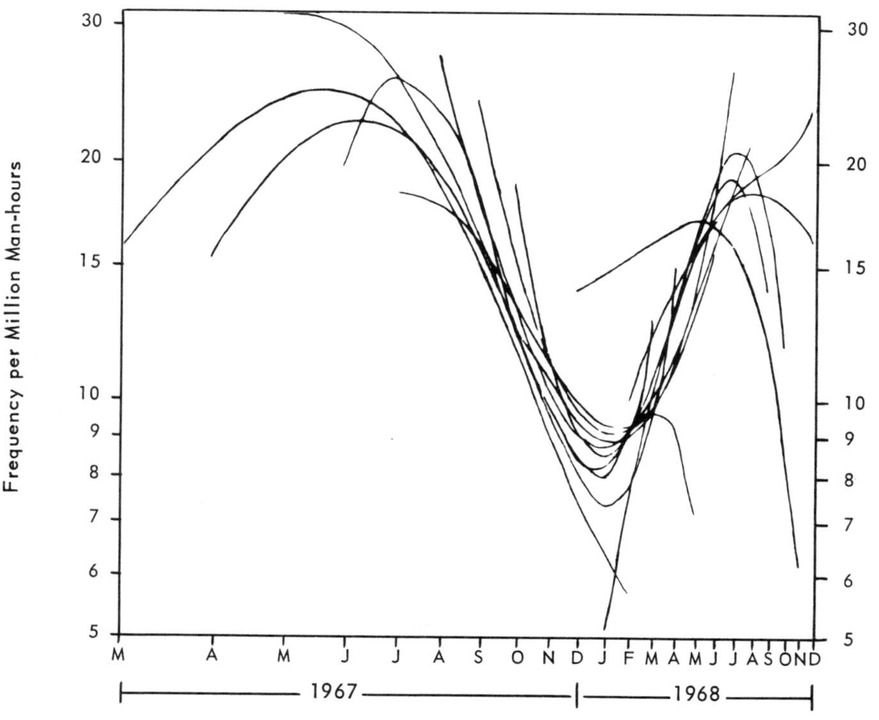

shows the composite curves for TIF. It should be noted that the SIF curves are more volatile, primarily because of the numerical effect of one injury at these levels of frequency.

The value of these short-term, 12-month MLR computations lies in the fact that changes are more apparent. The weight of all the previous years is not acting to moderate the tail, so earlier indications of change are available. During periods of poorer performance, as indicated by other injury-based index techniques, SIF tended to go out of phase with both the expected repetitive swings (seasonal) and TIF curves. This was experienced twice—once, in 1962, SIF simply had a significantly higher amplitude than was indicated by TIF; and again, in 1965, SIF increased while TIF was decreasing and, instead of tracking as is the case during periods of improving performance, SIF was simply out of phase with TIF. Both of these data gymnastics are indicators of worsening performance. As the curves tend toward the normal pattern, this improvement of the short-term situation is indicated.

Exhibit 29 has been included to show these periods. Certain techniques used to handle the data for this plot deserve explanation.

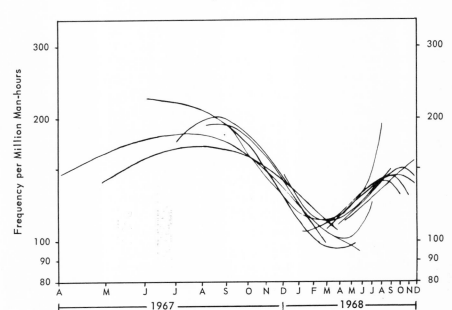

Exhibit 28
TIF Trend Profile Using 12-Month Learning Curves

Exhibit 29
A Performance Profile Using Short-term Learning Curve Averaging

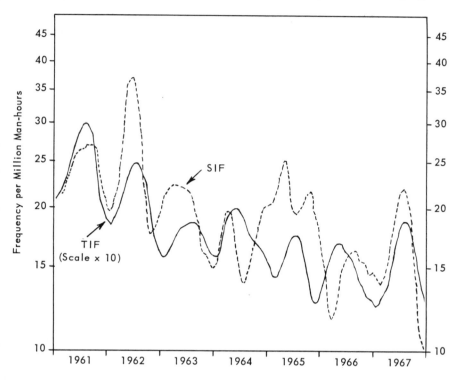

The time ordinate was not the log of man-hours but rather the linear ordinate by months. This was done to keep the curve stretched out rather than have it compressed as time passed. This makes the curve easier to read. The other technique was to use the average numerical value for each month. As stated before, the 12 points for any given month obtained from the various MLR computations were very close in value. To make the plot more readable, the average of these 12 values for a given month was used.

It is recognized that such handling is tedious and expensive. The study confirmed previous findings and, in so doing, validated the simpler methods. The one pertinent conclusion of this study was that the SIF and TIF learning curves calculated by the multiple linear regression technique served as a sensitive indicator of trend through the observation of the current end of the curve. The use of all available data in the development of the learning curve gives a general performance curve from which both the long-term trend and the effect of current performance are available. The former gives the basis for goal setting and shows what performance has really been. The little upsets that occur along the way, the crises of the moment, are washed out in this learning curve so the actual past performance trends are apparent and meaningful. The effect of current performance is pronounced and useful. Each new piece of data exerts a strong influence on the placement of the final regression point. It is both the position (value) of this final point and the direction (slope) of the tail end of the regression curve that provide the indicator of performance. In Exhibit 17, the goal line obtained from the extension of the long-term performance trend was shown with actual SIF data plotted to show how each month's performance was related to the goal. An improvement on this is now available through the use of the regression indicator. Exhibit 30 shows the same performance period and the same goal line, but the final regression values are plotted for each month, and an arrow shows the slope of the last quarter (last three values) of the learning curve. This requires that the MLR be performed each month, which, in turn, requires a computerized program. Where such computer equipment and programs are available, this data presentation is easier and more informative than the plots shown in the series of Figures 18 through 26.

Again, if the arrows have had a general negative slope over a period of months and are flattening out (less negative slope or even horizontal), a change is taking place. After three consecutive such points of change, we should be aware of the shift. After five consecutive such points of change, it can be concluded that a change has been caused and action to define this cause is in order. Of course, improvement can be recognized by the less positive or more negative slopes following a period of questionable performance.

The use of recent data (12 months or so) describes the short-term performance and has more volatile swings. If performance is being followed monthly by the long-term learning curve, an upswing or poorer performance could be confirmed by the learning curve MLR program on a shorter period of time with the most recent data. This is an expensive refinement that is hardly justified. The treatment of the data becomes more sophisticated than the data.

If any conclusions of the various work described in this chapter

Exhibit 30
A Current Performance Indicator Using Learning Curve Techniques

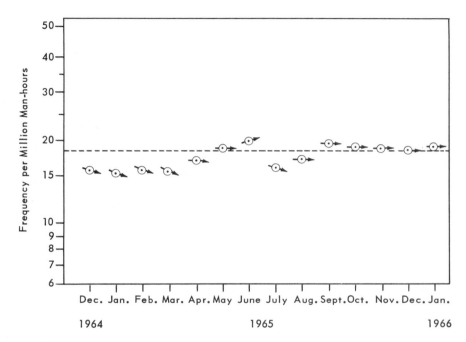

are to be repeated for emphasis and recommended to the reader for consideration, they would have to be these:

1. For the safety engineer in a plant in which computers are a standard tool of the business, the learning curve with multiple linear regression is an excellent method of presenting performance. It gives a profile of past performance and a practical goal for future effort.

2. For the safety engineer who finds time hard to come by and assistance unavailable, the cumulative sum methods, either of actual SIF and TIF data or rolling averages of these data, are easy to handle manually and provide the basis for understandable and effective presentation of safety performance trends to management.

3. For all safety engineers who seek to provide management with the tools for decisive control, the control chart methods described in Chapter Six are recommended.

CHAPTER EIGHT

THE POTENTIAL OR NEAR-MISS ACCIDENT

THE near-miss accident is one in which there is an event or incident, but because no loss occurs, there is usually no report. With the probability relationship of frequency increasing as severity of loss decreases, it is logical to say that the near misses probably occur often but are seldom reported. Although this may be understandable and accepted, it is a great loss to accident prevention personnel because it denies them the opportunity to investigate these incidents. In fact, it is probably true that when we investigate an incident which has resulted in injury, we are sampling the near-miss incidents. Tarrants, through the use of his Critical Incident Technique, has shown that most injuries are the results of incidents that were experienced, often repeatedly, in a work group without loss or injury.

Sampling

Since incidents do have causes but not all incidents result in losses, sampling for accident causes should generate data on which performance can be measured. Sampling methods and the statistical handling of such data are well documented in many textbooks, and such guides should be used by the safety engineer.

Some sampling method is in order to provide another indicator of performance. But what should be sampled? And what are some of the problems associated with safety sampling?

Generally the sampling system should cover a fairly wide spectrum of activity. It should not be limited to housekeeping problems, unsafe conditions, unsafe acts, and safety rule violations but should include the property loss potential incidents of spills, fires, vehicle accidents, activation of interlocks or other safety protective devices, and any other activity that had a cause and an incident but luckily did not result in a loss.

This would be a big order except for one very simple fact. Sampling, by definition, means to take a peek and, on the basis of the peek, form a whole picture. Most managers, if faced with a proposed program to sample everything, would simply throw up their hands and do nothing. So would most safety engineers. The best approach is to take it bit by bit. It is like a bottle of pills—if you take all the pills at one time you might die, but a pill taken as prescribed might make you well. A few categories should be selected, and the sampling program be established carefully to avoid the pitfalls that might make the whole effort ineffective.

Some Uses of Inspection

A housekeeping inspection is a qualitative measure of work conditions and worker attitude that lends itself to exception sampling. A housekeeping inspection program that takes a lot of time and generates long lists of temporary situations to be corrected discourages both the inspector and the poor fellow who gets hung with the laundry list. Start with a rather short list of housekeeping items that the inspector will look for in a defined geographic area of the business operation. Assuming that the standard of housekeeping that the owner or plant manager expects is "excellent" (not always what he really wants or is willing to pay for), then the object of the inspection is to find exceptions to the standard within a certain period of time.

The point of inspection is not to see how many housekeeping infractions the inspector can find so that a long list of items can be generated for correction. Nor is it to stay in the area long enough to find one or more infractions of each item. The purpose of inspection is to determine the degree of excellence of an area's housekeeping by seeing how many single infractions are observed within a fixed time period. For example, assume that Area A is to be inspected. The inspector, with a list of ten items, is to be in the area for fifteen minutes. He moves through the area at a crisp pace because the inspection time should be based on the ability of a given inspector to rapidly cover the area in one pass. This does not allow the inspector to hunt, but it does allow him to see a sample of the total area. When he observes an item that is listed on the card, he checks that item off the list and forgets about it. This is done because, first, he is looking for excellence and one deviation in one item signifies that conditions are not "excellent," and, second, the number of different deviations from the standard is itself indicative of the severity of the problems concerning housekeeping, working conditions, and work attitudes. A simple zero-to-ten rating applied to each inspection will develop data indicative of the housekeeping effort.

It is recognized that this is a rigorous inspection for excellence but not a very effective device for finding all the housekeeping items that need to be corrected in an area. But this is the character of sampling; it doesn't take a chunk of some busy manager's time, and it encourages excellence in one phase of the loss prevention program.

A second category for sampling could be the habits of the employees, as noted by the number of times they are observed performing work unsafely or violating safety rules. This type of sampling tends to be only qualitative, because workers who know they are being observed are more careful to work safely and obey rules during the observation. So sampling might best be done by those whose normal assignment brings them in contact with the work group. A supervisor who is really interested in the safety of his people can effectively do this sampling himself. He can purchase a small counter to carry in his pocket. Each time he sees and corrects one of his employees, he records that contact on the counter. At the end of each day or week, he simply reads his counter and compares present with past performance.

The manager can get a qualitative answer by applying the same technique used for housekeeping sampling. A list should be unnecessary, as managers or upper-level supervisors should know an unsafe act or rule violation without a list. However, a knowledgeable person may find, to his dismay, that these violations of good safety practices are committed repeatedly in the presence of the supervisor.

The Inspectors

The selection of inspectors is important. They should be people who as least understand the standard and are willing to have it applied in their own area of operation. However, they often have to be taught how to observe. One very effective way of training inspectors is to use slides or movies showing the infractions they are to look for. As the training progresses, it should become increasingly difficult to detect the infractions shown in the pictures. The trainer should flash the slide on the screen for ten seconds and then ask them to record the deviations.

It is necessary to make these sampling observations frequently so that the data have significance. When a group of inspectors is selected, either as individuals or in groups of not more than three, they should inspect areas on a random schedule each week. The data obtained should then be charted or tabulated for each area and trends noted. Again, areas should not be compared, but rather each area should be compared against its own previous performance. Because this is a qualitative measure, the actual ratings indicate little more than a general level of performance against the standard, but the trends are important.

Quantitative Answers—Property or Dollar Loss

A category that produces quantitative answers is that of the reported incident of accidents of small loss, usually property or dollar loss. In almost any business, there is a quantitative indicator of incidents that, in themselves or for other reasons, are not considered important enough to be formally reported but for which a report is nevertheless made because of a loss. For example, in a chemical plant or oil refinery, standard emergency equipment is provided in the operating areas. Included in this equipment are hand fire extinguishers. A record of those that have to be recharged and what areas they service gives a pretty clear picture of where small emergencies are occurring. The word "emergencies" was purposely used because it could mean a small fire, which wasn't serious enough to require the fire department, or an incident that occurred as a result of some thoughtless horseplay—a different, but serious, problem. Either of these reasons for using a hand fire extinguisher is an indicator of more serious incidents to follow.

Take another example. In almost every industrial operation, there is a system for writing work orders and accounting for their cost. Of especial interest to the safety engineer are those work orders that are written because something breaks down or someone "goofs," thus requiring work to be done at once on a high priority. In most cases, the emergency work order is indicative of a loss incident or poor work performance. In a department that has only an occasional emergency work order, the supervisor has the situation under control and the work is planned. It is usually also true of his safety effort. He has control of his people and his machines, and his safety performance will not have emergencies either.

In a department that has a large number of emergency work orders, it cannot be assumed that the safety performance is poor. There may be reasons for the situation beyond the full control of the supervisor. For instance, the design of the equipment may be inadequate, or the production schedule may be beyond the capacity of the men and machines. Or the boss may not care about safety, and the supervisor therefore finds himself in conflict with the apparent operating objectives. It may be that the work order system is such that to get anything done the supervisor has to claim that the work is an emergency. Perhaps the supervisor is just not very good at his job. In any case, a problem exists in any business or work group where the repair work frequently has to be done on an emergency basis. Whether it is directly related to safety or not, safety performance will suffer. For this reason, a running record of the frequency and cost of emergency work orders in an operating department or a plant is an indicator of performance and, sooner than one would think, of safety performance.

This chapter has dealt with ideas and concepts of safety performance measurement that are not as statistically fixed as those discussed in previous chapters. This does not mean that they or similar ideas do not have value. On the contrary, they form the basis for the judgments needed to make appropriate decisions concerning performance. Too often, in this computerized age, we become so engrossed in numbers that we lose sight of the limitations of those same numbers. Safety engineers must deal with the mix of men, materials, and machines. Even though safety engineers may be able to control the materials and the machines, men are less predictable. In fact, that is what makes them men—the ability to perform better than anyone should expect and to improve by learning from their failures. The safety engineer cannot depend solely on the statistical record of performance: It should be used to augment his other more qualitative measures such as housekeeping, morale, work habits, and so on so that the appraisal of safety performance is accurate and decisions made on the basis of this appraisal are right for their business or plant.

CHAPTER NINE

THE PROPERTY AND BUSINESS INTERRUPTION LOSSES

THE prevention of property and business interruption losses is just as much a part of the loss prevention program as the problem of personal injury losses. Unfortunately, the incident can result in losses to both property and people, but for the most part property losses do not involve personal injury.

The reporting of property or business interruption losses differs from the reporting of personal injuries. The injured man or the individual giving assistance can usually report the incident; the injured equipment cannot. Except for fires, explosions, or major equipment failures—which, by their dramatic nature, make the incident known to many people—most losses are handled without fanfare and are investigated internally within some segment of the organization.

The Reporting Situation

The report itself differs in scope from the injury report. Usually the injury case involves one person within a small group. The causes and effects are limited to that individual and that small group. Property loss incidents usually involve actual or potential causes and effects beyond a minor loss or the loss of a single piece of equipment.

This is especially so in view of the increasing complexity and interdependencies of industry. Since most of the required information concerning a property loss incident must be obtained through an investigation, the investigator who has to report the information is the critical person in the loss control effort. Whether the responsible supervisor or a committee performs the investigation, it is imperative that the safety engineer receive a copy of the report. The act of reporting must have top management encouragement. Although immunity from accountability is impossible, the supervisor must know that management also recognizes his performance when he investigates, corrects causes, and reports incidents. This paradox of accountability probably explains the fact that relatively few reports are usually obtained.

From the safety engineer's point of view, this entire reporting situation is frustrating. He, like the accountant or the technical service engineer, is in a staff position to assist the supervisor with his problems. But the results are different. What supervisor would—or even could—keep the cost figures on his operation from the accountant? There are very elaborate procedures, developed and accepted long ago, that make the cost figures available to the accountant. From this information, cost statements are prepared for the manager. But these same cost statements are a tool used by the supervisor to measure and control costs. The supervisor with a technical problem provides whatever information is needed by the technical services engineer in the hope that the problem can be solved. The supervisor accepts the fact that, in requesting assistance, he is notifying management that he has a problem. But when it comes to loss prevention, the same supervisor avoids contact with the safety engineer because he would prefer not to advertise his safety problems.

It would seem that costs and technical problems are business situations to be approached and analyzed dispassionately so that a logical business decision can be reached. However, loss prevention problems have emotional overtones and require different approaches. At least part of the blame for this attitude lies with those safety engineers who have promoted the philosophy that all incidents are bad and anything that *can* happen *will* happen. And a part of the blame lies with managers who cannot or will not understand that loss problems are both cost problems and technical problems, that **they** have causes and effects, and that responsible people with professional expertise are needed to define and to solve them. When the safety engineer demonstrates his expertise, and when management recognizes that the solution is a means to protect profits, loss incidents will be readily reported to the safety engineer and a new dimension in loss control will be established.

The Property Loss Incident Form

The report should be in a convenient, understandable form so that its format or complexity does not discourage reporting. A sample report form is shown in Exhibit 31.

The front of the form allows the investigator to identify the subdivision of the busines operation in which the incident occurred, when and to what equipment it occurred, and to describe what happened and why. The investigator also enumerates the value of the loss as property and/or business interruption. A list of categorized causes permits an easy check of those that are applicable. Finally, a space is provided for the investigator to state the corrective action and by whom and when it is to be completed.

The back of the report form collects information for analysis. The upper half, to be completed by the investigator, concerns the potential of the incident. Sometimes this may be the loss actually experienced and recorded on the front, but often it is considerably greater than the actual loss because the loss sequence has been interrupted by the quick action of personnel, activation of protective devices, or other circumstances. The *probable maximum loss* is determined and recorded. The probable maximum loss is defined as the loss that probably would not be exceeded, even with full realization of the loss capability. An example of the difference between the actual and potential loss can be given in an incident of a pipe rupture of flammable gas within a building. If the gas does not ignite, the actual loss would probably be limited to the cost of repairing the pipe and the loss of production while the repairs were made. But if the gas ignites with explosive force, the building might be destroyed, with extensive equipment loss, business interruption, and personal injury. The probable maximum loss would be considerably greater than the actual loss. However, this estimate of the probable loss would not be reasonable if equipment and buildings within a mile radius were considered lost when in fact the explosion effects were limited to the building.

The basis for this estimate of probable maximum loss must be explained so that others not directly involved in the investigation can understand how the potential was evaluated. Further, the planned action based on this estimate of loss may involve long-range plans or engineering studies that are justified by the potential but not by the actual loss or that appear unnecessary in light of the actual loss.

On the lower half of the back of the report, the risk analyzer classifies the incident and records the relative probability of the incident. (Risk analysis will be covered in detail later.) The determination of the probable maximum loss and the relative probability provides a priority for this and similar incident causes and a justifiable basis for the effort for corrective action.

The reports of property loss or business interruption incidents have value in the loss prevention effort, but unless a sizable number are obtained, their value as measurements of performance is limited. Like the major or disabling injury loss, the major property loss is probably dramatic enough to generate the needed reporting conditions. A major business interruption incident might not be reported if the property loss resulting from the same incident is minor. A moderately serious property loss or business interruption, unlike the serious injury, may not be reported. And minor property losses are less likely to be reported than minor injuries. It is not that these losses are never reported but rather that they are not reported to the loss prevention engineer who could use the information to assist in problem definition and to alert other supervisors by making the pertinent information and analyses available to them. In most industrial plant organizations, there is a procedure by which the manager receives daily reports from the various operating subdivisions that recount the events of unusual occurrence of the previous day. Many of these reported occurrences are, in fact, losses but are accepted as operational upsets rather than reported as actual or potential losses. This source of data represents a valuable insight into the plant performance but may not provide a base for measurement in the statistical sense.

Exhibit 31
The Property Loss Incident Form

REPORT OF PROPERTY LOSS INCIDENT	**NO.**	Page 1 of 2

Use this form to report events which are sudden, unscheduled, and which result in, or have the capability to result in, property/BI losses over $500 and/or serious or greater personnel injury. Ref. P-0111

Distribution (By LP & S Section)
Plt. Mgrs. Safety Board

Dept.: _____ Product/Function: _____ Date: _____ Time: _____
Functional Item No.: _____ Name: _____
And/or Area: _____

DESCRIPTION OF INCIDENT: Summarize completely in this space. Details may be covered by attachment if necessary.

INJURIES: ___ None ___ Minor ___ Serious ___ Major ___ Multiple Major

LOSSES: Bldgs./Equip. $ _____
Stored Raw Matl/Product $ _____
BI (Profit & Cont. Exp.) $ _____
TOTAL LOSSES $ _____

FUNCTION with which the incident was associated: ___ Mfg./Utilities ___ Maintenance ___ Services ___ Construct./Demolition ___ Lab/Pilot Plt. ___ Other:

PERSONS DIRECTLY involved: ___ None ___ Monsanto Employee(s) ___ Non-Monsanto: ___ Routinely in Plant ___ Not Routinely in Plant ___ Outside Plant

CAUSES: (Check those applicable):
___ Operating/Procedural Error
___ Questionable Process/Control Design
___ Questionable Mech./Electr. Design
___ Defective Equipment/Component
___ Installation or Maint. Error/Omission
___ Corrosion/Erosion/Fatigue/Wear
___ Instrument/Control Malfunction
___ Lack Routine Surveillance/Servicing
___ Lack Reasonable Preventive Maint.
___ Natural Phenomena

CORRECTIVE ACTION: What is being done to prevent a recurrence of this incident? Responsibility? Completion date?

Form Revised 7-68

Forward to LP & S within 5 Days

(FRONT)

Exhibit 31
The Property Loss Incident Form (continued)

ANALYSIS	INCIDENT NO.	Page 2 of 2

PROBABLE MAXIMUM LOSS: Estimate the maximum loss which might reasonably have resulted out of this incident: (Or Check: _____ Same as Actual)

Total Property/BI: _____ $0-500 _____ 500-25M _____ 25M-100M _____ 100M-1M ☐ _____ Over 1M ☐

Injury: _____ None _____ Minor _____ Serious _____ Major _____ Multi.Major

BASIS: Describe your basis for this estimate: (Explosion, Fire, Destructive Breakdown, Etc. Extent of damages, percent and duration of BI disability, etc.)

PREVENTIVE ACTION: What action is planned to avoid this PML? Include both immediate and long-range plans, studies, assignments, etc. Responsibilities? Completion dates?
(Or Check: _____ PML does not warrent further action.)

Reported By: _____ Date: _____ Reviewed By: _____ Date: _____

LP & S REVIEW: Classification of Loss Incident Eo: ☐

_____ Loss Proc/Mfg Contr. _____ Pri. Failure, Cont Equip _____ Pri. Failure, Electr Equip. _____ Primary Ign.

_____ Loss Contr.-Other _____ Pri. Failure, Mech Equip _____ Loss Essential Input _____ Ex-Design Event

LP & S Review By _____ Date _____

(BACK)

The Work Order as an Indicator

Most businesses with real property and equipment have a system of work orders, a paper request for work worth some nominal amount of dollars. In the procedural steps of such systems, this piece of paper is authorized by the supervisor operating the equipment, indicating his approval for the work to be done. The work order authorizes the supervisor responsible for repairs to obtain the manpower and perform the work. Once the work has been completed, the costs of this job are collected, possibly by an accounting department, and the operating department is charged for the work. In such a system, there is some priority assigned to various work orders to differentiate between those jobs which must be done on an emergency basis and those jobs which are less pressing and can wait until manpower is available.

A measure of property loss performance is available from the number and cost of emergency work orders. In effect, these are the minor injuries of property loss because they constitute the unexpected and unplanned property activity in the plant or business. Of course, emergency work orders may be written on some work not related to a property loss (just as some alleged injuries didn't occur in the course of employment). Also, emergency work orders may be written to beat the system in order to get manpower that would otherwise not be available. As in the case of the minor injury, the data will not be entirely accurate, but the probability relationship will exist and the data are still significant. Cumulative sums or control chart techniques can be used to show trends to supervision and management.

As discussed in earlier chapters, the statistic used to present personal injury performance trends is injury frequency. In the property loss area, the parallel statistic is the property loss frequency. This could be as simple as the number of emergency orders issued per month.

$$\text{PLF} = \frac{E}{T}$$

where PLF is the property loss frequency.
        E is the number of emergency work orders.
        T is the period of time, say, a month.

This does not really describe the loss frequency because it ignores the variability of exposure. This exposure could be stated as the percent of rated capacity of the operation, since as the operating throughput increases, the occurrence of a loss incident becomes less desirable and more expensive. The incident interrupts the plan to attain the throughput goal and tends to be more of an emergency than a similar incident would have been considered at lower production rates. Also, the incident carries a higher probability of reaching the loss capability during periods of capacity operation. A warehouse which is only half full does not exhibit as high a loss potential as a full warehouse, assuming the contents are of similar value. A continuous oil refinery unit operating at full capacity to meet customer demand has a higher exposure and probability for the emergency event than one operating at reduced rates. Therefore, the index for exposure measurement might be expressed as the number of emergency work orders per level of operating capacity.

$$\text{PLF} = \frac{E}{PT}$$

where P is percent of rated capacity,

or,
$$\text{PLF} = \frac{\Sigma \$ E}{CPT}$$

where $\Sigma \$ E$ is the total cost of emergency work.
    C is the total invested capital of the operation,

or,
$$\text{PLF} = \frac{\Sigma \$ E}{(\Sigma \$ M)T}$$

where $\Sigma \$ M$ is the total cost of all repair or maintenance work.

Any of these property loss frequencies describes the level of emergency work activity during a given time period compared to the total activity of the operation. Although these may not be true and accurate descriptions of the loss exposure frequency, it is the trend that is important. Any expression in which the number or dollars of emergency minor work are equated with time periods or activity level during time periods will give a trend picture of these minor losses. Again, comparison with unrelated businesses or even the same business across the street will lead to inaccurate conclusions about performance.

An advantage of these work orders or the dollar costs as performance measurement indexes is that the supervisor, under this pressure, tends to adjust his ideas of priority. He will reduce the use of the emergency work order as a means to beat the system, and he will improve his planning of the work. He will pay more attention to preventive maintenance. All these actions will result in cost improvements and loss prevention.

Of course, the major property loss or business interruption may not be directly prevented by performance measurements and controls based on emergency work order data. But the attention to detail and awareness of the loss potential will create the attitudes and discipline necessary to be aware of and to correct conditions that might lead to major losses.

The personal injury is a loss, just as the breakdown of equipment or the stopping of production is. There is a probability relationship between minor injuries and serious or disabling injuries. There is also a probability relationship between minor property losses and serious or disabling plant losses. But until there is a program for performance measurement to give the direction of need, managers will continue to respond only to the major losses, and correction will produce only partial loss control. Through the use of loss trend or control charts, the supervisor and manager can be given basic information about performance and can develop the necessary controls to improve that performance.

CHAPTER TEN

THE LOSS CONTROL PROGRAM

THE loss control program is the directed effort to recognize, evaluate, and correct loss exposures. It is the activity that prevents loss and protects people and profits. It is *management's* program. The safety engineer can recommend the concepts, promote their use, plead for their acceptance, and participate in the resulting program; but without management's understanding, acceptance, and leadership, the loss control program cannot be effective.

Loss Control Theory

There is an axiom which must precede any discussion or understanding of loss control: All losses cannot be prevented without ceasing to produce the product and to make a profit. Production and profits depend upon the physical and mental efforts of people using various materials and a wide assortment of equipment and machinery to form a product. These people, materials, and machinery have value, and the loss of any is a loss to the business. Whenever there exists the capability of a loss, there must also exist the mathematical probability that the loss will occur. The losses will occur at a frequency dictated by this mathematical probability. This is not to say that if a loss can occur, there is nothing we can do about it. Nor is it to say that if a loss is possible, the preventive action is demanded. The first attitude is fatalistic and the second is unjustified. Incidents are probabilistic in nature, and only if the probability and the potential threaten the profitability of the business are preventive measures justified. In order to recognize a threat, we must determine and understand the probabilities and the potentials of loss incidents. "Loss prevention" is not really prevention; it is *loss control*. And loss control is the art of attaining the optimum balance of loss potential, loss probability, and profit.

Over the years, there have been many approaches toward the development of a loss control program. The obvious approach was the investigation of accidents to identify causes and provide correction. This approach depends on the occurrence of the loss incident, usually one that resulted in a physical loss either to people or to property, and it is an effective way of reducing similar incidents in the future. However, its effectiveness is limited by two factors: First, it cannot be used until an incident occurs, and, second, the lessons learned must be communicated to and recognized by those who have similar exposures. Examples of the success of this approach have been the boiler and machinery insurance organizations' contribution to boiler standards, inspections, and protective devices; American Petroleum Institute standards for the oil and petrochemical industries; and National Fire Protection Association standards for fire prevention and control. In addition, the Manufacturer's Chemists Association has provided data sheets on the handling of chemicals and incident reports; the National Safety Council has developed and distributed quantities of quality safety information and program material; the United States of America Standards Institute has instituted general safety standards. Many other organizations and individuals have made incident information available and called our attention to the safety lessons to be learned from often painful experience.

The Business Process Safety Review

The preventive effort was not always content to wait until a business was built and an incident occurred. In response to pressure, usually initiated by some near-disastrous event, the Business Process Safety Review became a part of the loss prevention program. Although this review or audit has taken many forms, it is fundamentally an investigation of the business process to identify potentially hazardous conditions or failures of the design and to make such alterations as are deemed necessary in order to adequately protect people and property. In general, the review deals with spcific areas of concern, such as fire protection requirements, utility reliability, and physical separation of process material storage. The normal procedure is to study the facility in detail, methodically going through the process step by step. The knowledge and experience of the research, design, manufacturing, and marketing personnel, coupled with the inquisitiveness of the safety engineer who is armed with organized information, can result in an effective safety examination of a new or existing business process.

One result of this effort in loss prevention is the availability of detailed lists of items, which insures that various areas of concern will be examined thoroughly and in great depth. Examples of two types of review aids are, first, the *Guide to Fire Prevention in the Chemical Industry* and the *Hazard Survey of the Chemical and Allied Industries*. These cover a wide scope of subjects from site preparation to applicable legal codes. Second, "Hazard Classification and Protection" in the *Safety and Loss Prevention Guide* develops a fire and explosion index by assigning quantitative values to determined hazards and suggests protective features to be incorporated in the process facility for various values of the index.

Reviews and examinations of new facilities must find loss exposures and provide for their correction early in the design, before the physical design and the money are irrevocably committed. This can effectively reduce the cost of a project, because changes after equipment has been ordered or installed can be extremely costly. The review will indicate a specific area of design revision, provide the justification for change, and, when corrections are made, improve the reliability of the process.

This kind of review of both existing and proposed processes is accepted by engineers because it begins to approach a scientific method for loss prevention. It is accepted by management because the action requested is justified. It can develop a priority for realistically defined problems and call for action based on anticipated need. It highlights omissions and duplications, so that both can be corrected. It builds confidence in the process because the problems have been recognized, studied, and corrected to prevent incidents and losses.

The Business Process Safety Review should be performed on all new process designs, all revisions of existing facilities where the conditions of operation are changed or new materials are introduced, and all test programs resulting in temporary significant changes of the operating conditions. The review should be initiated by the individual or group requesting the change or performing the design. Managers can be assured that the review has been performed by requiring verification of the review before approving the expenditure of funds.

Systems Safety Analysis

Accidents, at whatever level, result from systems failures. The concept that accidents are caused by unsafe conditions, unsafe acts, or both is really just another way of stating that the systems—which consist of man, machines, and materials to produce a product or service—have failed. These systems are designed or planned to operate under control to produce a certain product or service. It is the loss of this control which results in losses—injuries to people, destruction of property, and interruption of business. Therefore, loss control can be attained through systems analysis and control.

There seems to be some controversy over the applicability and justification for systems analysis in conventional safety and loss prevention activities. The techniques of analysis used, for example, in the missile and aerospace programs are quite formal, elaborate, and expensive. Although systems safety analysis in business does not need to be either as complex or as costly as that applied in aerospace, people in business apparently regard such programs as beyond the capability of their organizations and analysis has not had a fair trial in the industrial or business loss prevention programs.

Nonetheless, in recent years various researchers have made significant contributions toward bringing systems safety analysis into the engineering discipline of loss prevention in industry. Of these contributions, only three are mentioned here to indicate the diversity of the areas of work; it is not intended to overlook others nor to imply that these are the only contributions in this field.

Louis B. Kahn provides a scholarly "technique for determining a finite probability of destructive failure of a plant producing a hazardous product."[1] He develops the probability of a plant failure related to the reliability of the operating and safety systems for an uncomplicated plant.

G. H. C. Eng assesses the adequacy of alarm and shutdown systems: Will they perform as required during a fault situation? The assessment of the accuracy, response, and reliability of a given safety system can indicate the capability of the system to safely shut down the plant. Such analysis will lead to the recognition of weak points or redundancy in the system so that these can be corrected.

George A. Peters and Frank S. Hall speak of three methods of performing a design safety analysis. The first is a review of the design by a group that is organizationally and functionally independent of the designers. The second is an analysis for catastrophic failures arising from "all probable failure modes, their causes and consequences."[2] The third method is a design hazard analysis, a new technique to apply systems analysis to the design. It results in a quantitative hazard analysis matrix.

It is the purpose of this chapter to suggest a method of systems review which will identify the loss exposures, analyze the causal factors, and develop the potentials and probabilities of the undesired loss incident.

If control is to be realized, we must be able to define, develop, and quantify the valid loss exposure, its potential, and its probability. This requires formal analysis of the loss process—quantifying the probabilities of loss inputs and outputs and measuring these findings against the tradeoff yardstick to determine whether the exposures meet established standards. This is the job of the safety engineer. The manager's job is to decide whether the correction of the loss exposure is justified. He cannot make a logical decision until he knows what the true loss exposures are, whether the standards have

---

[1] Louis B. Kahn, "A Statistical Model for Evaluating the Reliability of Safety Systems for Plants Manufacturing Hazardous Products," *Technometrics*, August 1959, Vol. 1, No. 3, pp. 293–307.

[2] G. A. Peters and F. S. Hall, "Systems Safety Engineering as a Technical Discipline," paper presented at the Ergonomics and Aerospace Joint Session of the American Industrial Hygiene Conference, April 26–30, 1964, in Philadelphia, Pa.

134   ACCIDENT PREVENTION AND LOSS CONTROL

been met, and how much time and money will be required to correct them.

To define loss exposure, the most efficient and thorough method is systems safety analysis. The deterrent is not its complexity or its expense; it simply has not been put into a usable business form. The following section will attempt to develop systems safety analysis procedures that are applicable to industry and business.

Loss Analysis Diagram

To illustrate the loss process, a logic diagram is used. In the diagram, events and conditions will be shown as the outputs of other (input) events and conditions, and logic "gate" symbols will be used to indicate causal relationships. Because the simplified logic diagrams discussed in this book pertain particularly to analysis of the business loss process, the name "loss analysis diagram" (LAD) is used. The general diagram is shown in Exhibit 32.

Events and conditions are symbolically presented in LAD, with the ultimate loss given at the top of the diagram. This loss will be the probable maximum loss (PML) for the particular system being analyzed and is the loss potential for that system. The PML includes direct damage and business interruption costs. The direct damage

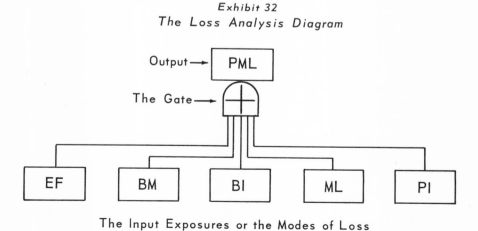

Exhibit 32
The Loss Analysis Diagram

The Input Exposures or the Modes of Loss

costs include the repairs or replacement of the property damaged or destroyed. The business interruption costs include the profit loss plus proportionate fixed or continuing expenses for the period of disability. The exposures or modes that could lead to the probable maximum loss can be identified either through the reporting system (described in Chapter 9) or by methodically studying the design or actual operation to determine what the PML would be in the event of a loss of this vessel, compressor, or operating system. It is obvious that some minimum dollar loss figure exists below which the effort of analysis is not justified, say, a PML of $100. On the other hand, a PML minimum of $10 million would scarcely provide any practice of the analysis since such losses, fortunately, are rare. If the business has a property insurance policy containing a deductible figure, this amount would provide a good minimum point, that is, the analysis of any exposure which could develop a PML equal to or greater than the deductible level of the property insurance policy. This may be an acceptable minimum PML because the insurance companies, through experience and negotiation, have found the lowest dollar point at which they can accept claims at a reasonable premium and still make a profit from insuring your business. Alternatively, you may arbitrarily set a minimum PML of $10,000, $50,000, or whatever. As correction of the exposures is completed, this minimum can be decreased, presenting more and more exposures for analysis.

PML is the output into which input events and conditions are fed. In the diagrams these inputs are connected to the output through logic gates, indicating cause-and-effect relationships.

The *or* gate signifies that *any one or more* of the various input conditions is necessary and sufficient to produce the output event or the loss. In Exhibit 33, the *or* gate shows that for output A to occur, it is necessary and sufficient that only one of the inputs—B or C or D or X—be satisfied. In the electrical analogy pictured, it is necessary and sufficient for only one of the switches to be closed to close the electrical circuit and energize the light bulb, A.

Exhibit 34 shows the *and* gate, which means that *all* the input conditions, B through X, must be present to produce the output event. The absence of one input is sufficient to thwart the output event. In the electrical analogy shown, the light bulb will be energized only if all the switches are closed—B and C and D and X.

Exhibit 33
The OR Gate

Diagram

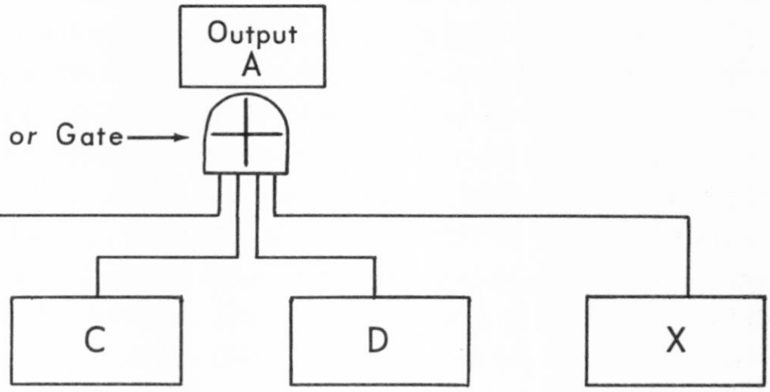

Output A

or Gate →

B C D X

Gate Circuit Analogy

B
C
D
X

A

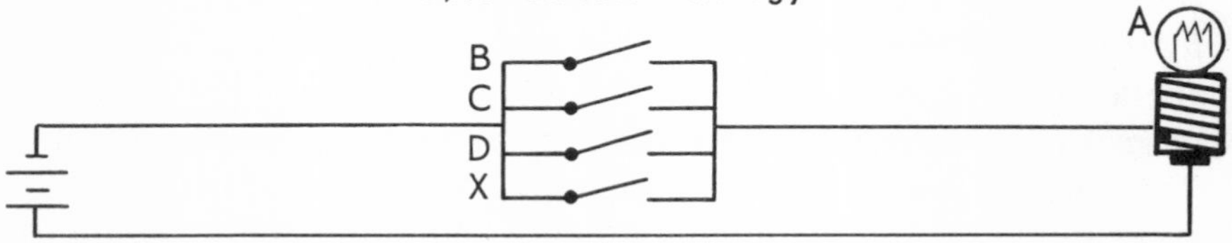

Exhibit 34
The AND Gate

Diagram

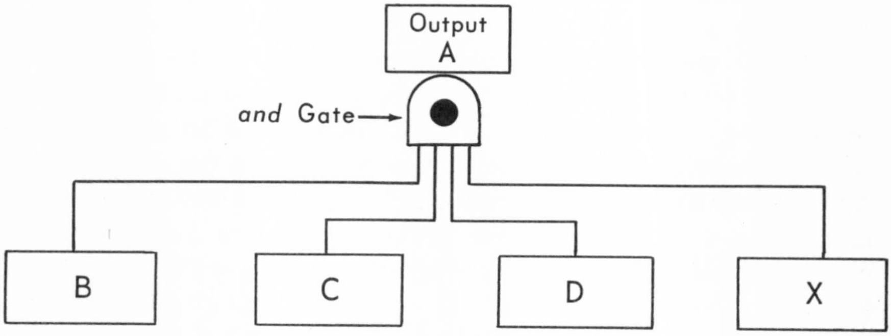

Output A

and Gate →

B C D X

Gate Circuit Analogy

B C D X

A

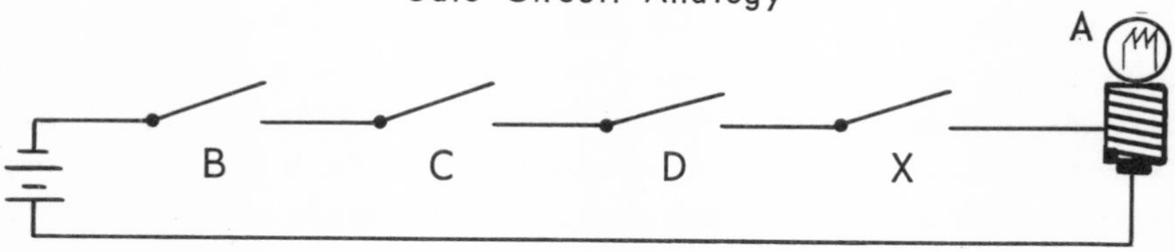

In systems safety analysis, the word "mode" refers to the manner or process by which the loss is sustained. These modes of loss are the inputs of the diagram in Exhibit 32. Destructive modes are those loss processes that result in damage or destruction to physical property and people. There are also nondestructive modes of loss that do not result in damage to property or people although they do involve the interruption of the business. In each loss mode, there must be a force capable of causing the loss, an object (property or person) acted upon by the force and sustaining the damage or interruption of activity, and finally a loss incident which is the sudden, unplanned event capable of resulting in the loss.

Destructive Modes of Loss

There are two basic destructive modes by which loss is sustained. The first of these is explosion–fire (EF), in which the damage is generally caused by the energy of chemical combination and reaction. The EF mode can be subdivided into three general classifications.

1. EF1, or primary fire, is a mode in which the fuel is routinely accessible to ignition without the necessity of a prior incident. An example would be the presence of a combustible material, such as paper bags or wooden pallets, in a warehouse. Only ignition is needed to change the loss exposure into a loss.

2. EF2, or secondary fire, is the mode in which the fuels are present but contained in pipelines or vessels and require a prior incident to make them accessible to ignition. An example of this mode would be a storage tank of gasoline. The gasoline may be safely contained, but it can cause and sustain loss if it is accidentally spilled from the tank. Once spilled, only ignition is needed to result in a loss.

3. EF3, or destructive reaction, is a mode in which the materials can destructively react if subjected to some physical condition outside the designed or normal conditions or if inadvertently combined with other but incompatible materials. Examples of this mode are runaway reactions, uncontrolled polymerizations, erroneous mixing of materials that causes vaporization, disassociation, or chemical reaction. The events must proceed to a condition beyond the design

capability of the equipment or container. One such incident often reported is the explosion of an oxygen or compressed air system as a result of oil in the system.

These three destructive modes—primary fire, secondary fire, and destructive reaction—describe the process by which damage is caused by reactive energy.

The second method by which loss is sustained in the destructive mode is mechanical in nature. In the language of the insurance industry, such losses are classified as boiler–machinery (BM) losses. The damage is attributable generally to nonreactive energies of motion, pressure, heat, and electricity—for example, broken shafts on moving equipment, overload of electrical circuits, and overspeed of a driver engine. These losses result from the misapplication of mechanical energy, exceeding the capability of the specific equipment. This is not a case of age or wearing out, because such failures are not loss incidents. These are *expected* events and do not meet the definition of being sudden, unplanned, and capable of loss. BM modes are those processes leading to loss that are outside the mechanical design of the specific equipment.

Nondestructive Modes of Loss

Let us now consider nondestructive modes of loss. The most common in industry is what is known as business interruption (BI)— the process of producing for a profit is interrupted even though there is no physical damage. There may be and often is a business interruption in the case of a destructive loss, but this is treated as a part of the total destructive loss. BI occurs with the loss of essential input (BI1), such as power, feed stock, cooling or heating media; or with the restraint of the process (BI2), such as line plugging or loss of process control. In this mode of loss, production is stopped or reduced without physical damage to equipment or change in the original design capability of the process.

Another nondestructive mode is material loss (ML) resulting from the inadvertent flaring, sewering, or dumping of a product or the degradation of materials used in the process. These losses are usually associated with a loss of process control and, again, are not necessarily accompanied by any physical damage.

A third nondestructive mode is the accidental injury of personnel (PI). Injury to personnel is not considered nondestructive, but this discussion concerns only the incident in which a person is injured and a loss does occur, perhaps only requiring a band-aid, but in which no physical property damage is sustained. Again, it should be noted that personnel injuries can (and often do with great severity) develop out of the destructive modes.

These definitions of modes for loss provide easy classifications of the manner in which the loss occurs.

Other symbols used in logic diagramming are rectangles, diamonds, and circles. In Exhibit 35, the output (the functional failure) is described by three failure inputs, any one of which is sufficient to cause the output failure. This figure illustrates the modes of failure (not loss) and is a basic diagram within the loss analysis diagram. The rectangles indicate events, inputs, or conditions, which must be explored further to develop other inputs necessary to the completion of the analysis. If the event is a failure, it will always occur in one of three ways.

1. The primary failure, designated by the circle, occurs while the equipment, process, or system is operating within the design conditions. This might come about if the equipment has deteriorated by corrosion, erosion, or fatigue; if the equipment had been inadequately designed; or if the equipment was defective in material or workmanship.

Exhibit 35
Functional Failure

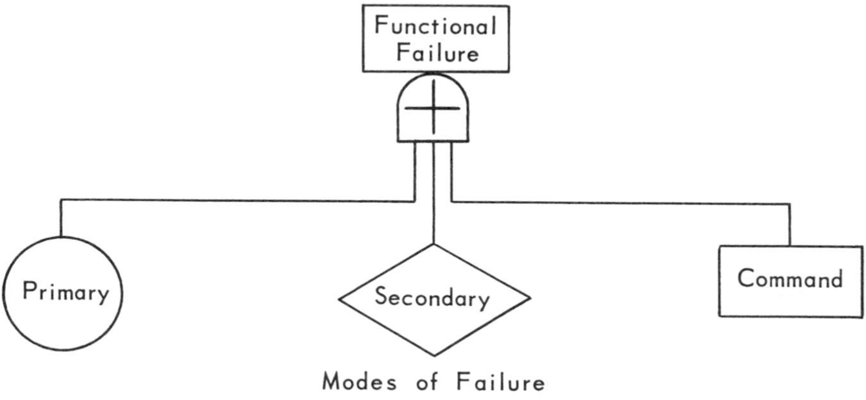

Modes of Failure

2. The secondary failure, designated by the diamond, occurs because of environmental or other external causes. The equipment may have been loaded beyond the level for which it was designed —for example, exposed to high winds, lightning, adjacent fire, explosion, or other external events. These causes are not usually deserving of further examination, and therefore the diamond is a termination symbol.

3. The command failure, designated by the rectangle, occurs because the system was instructed to fail. Here the equipment fails either functionally or physically because of various actions or interactions. Command failures are always investigated to develop the cause sequence.

Whatever the input events are, they must be both necessary and sufficient, in the context of the logic gate, to produce the output event. Otherwise the diagram is invalid.

EF LADS

Based on these principles for loss analysis diagramming, the loss processes will be illustrated by expanded diagrams. Each loss mode has a characteristic LAD form, and these generalized diagrams will assist the analyzer to visualize loss situations and to make preliminary appraisals of loss probability.

LAD for the EF1 (Exhibit 36). We will begin with the EF1 loss processes. The undesired loss is given in the rectangle at the top of the figure. Three input conditions are necessary to result in an EF1 loss. The first is the *exposure*, the capability to cause and to sustain loss. Both of these capabilities are satisfied in the presence of flammables routinely exposed to ignition and to something that will be damaged by its burning. If the flammables were removed, the loss could not occur. Since this is not always possible, it might appear that this exposure should be symbolized by a diamond, indicating a terminal point. However, the rectangle indicates that the exposure is not a constant and that the quantity of the fuel can be controlled, even though elimination of the exposure is not possible. For example, a warehouse is used normally to store flour packaged in paper bags and boxed for the marketplace. From time to time, the boxes and bags get damaged and the flour has to be repackaged. The empty boxes and bags are discarded in an unused corner of the warehouse to be picked up later and removed from the building. But as long as they remain in the building, they constitute an exposure capable of both causing and sustaining a significant loss in that warehouse. A housekeeping program might significantly affect this exposure level.

The second input condition is *ignition* resulting from the failure of such primary controls as no-smoking rules, nonsparking tools, hot work forbidden in the area until certain control measures are taken, or other restrictions used to control ignition.

The third condition necessary to the ultimate loss is a *breakdown* of the fire controls intended to function after the fact of ignition. In the warehouse example, the sprinkler system may fail to activate and the fire department may fail to secure the fire situation. Both of these protective devices are subject to primary, secondary, and command

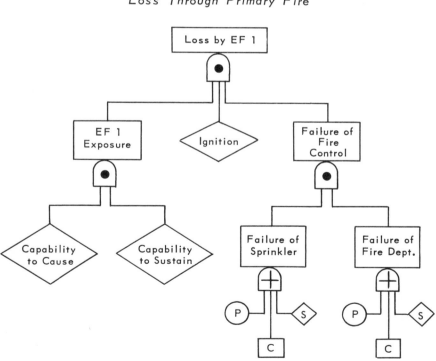

Exhibit 36
Loss Through Primary Fire

failures. If the sprinkler system fails to operate because the water control valve fails to open, this is a primary failure. If the sprinkler system cannot function because of some unrelated event, such as a ruptured water main, this is a secondary failure. The command failure occurs when some person shuts off the water to the sprinkler system so that, although it is capable of functioning as designed, it is "instructed" not to operate. Investigations of fire losses in which sprinkler systems did not function often show this command failure to have been the secondary control failure which permitted the loss to reach its ultimate. It should be noted that the failure of the fire controls is necessary if the ultimate loss is to occur. Therefore, it is necessary to satisfy the *and* gate. However, even when the fire

Exhibit 37
Loss Through Secondary Fire

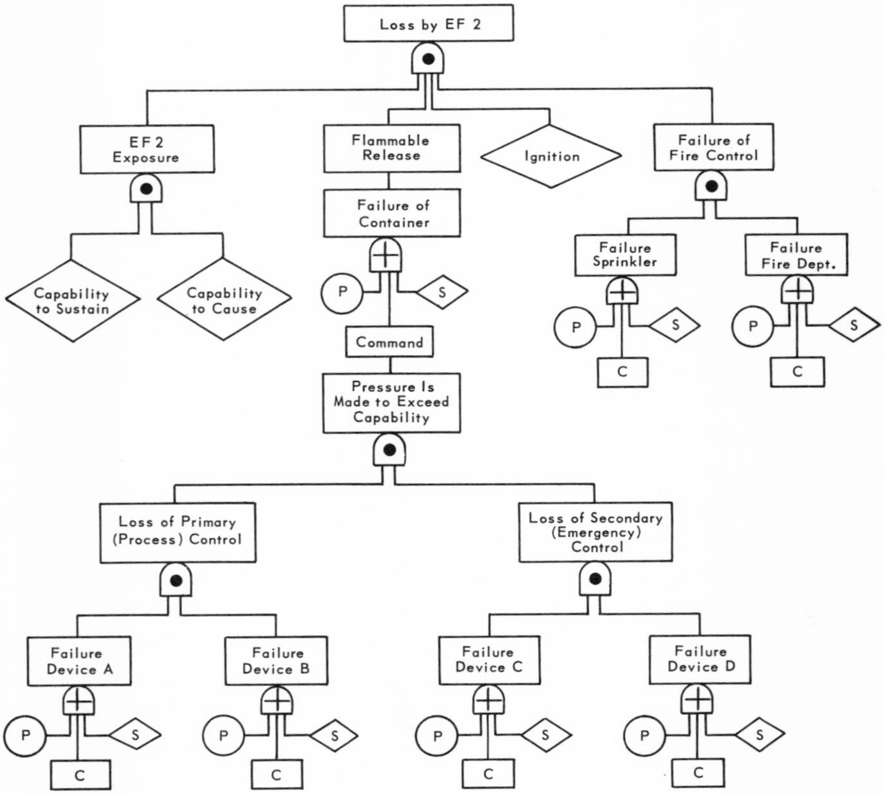

controls function successfully, a loss occurs, but the severity will be less than ultimate. This type of situation is discussed when the probabilities of failures are considered.

The LAD for the EF2 (Exhibit 37). The difference between the EF1 and EF2 modes is whereas the fuel is routinely present in EF1, it must be released by a prior incident to be accessible to ignition in EF2. The exposure is the contained fuel in EF2; ignition and failure of fire controls are similar to those of EF1. The release of the fuel from the container is necessary for the loss to occur. There are three circumstances in which the container will fail to contain the fuel. Primary failure is one in which the container capability is exceeded by normal demands, such as might occur as a result of corrosion of the container, defective fabrication, or improper design.

Secondary failure is caused by some outside event that brings stresses beyond the capability of the container to withstand. An example of such a failure is the rupturing of a storage tank by lightning.

Command failure is the mode most often experienced in losses stemming from the release of fuels. Here the containing conditions are changed, causing the capability to contain to be exceeded. In such systems, there are usually both primary (or process) controls to maintain the container conditions within design and secondary (or emergency) controls to protect the container if the primary controls fail. Therefore, for the command failure to take place, both the primary and secondary controls must be lost, as indicated by the *and* gate. The loss of primary control means that all the devices for maintaining control of the process are lost. Of course, each of these devices may fail in the primary, secondary, or command methods. The loss of secondary control may occur at any time but remain unknown until the primary control has failed. For the ultimate loss to be experienced, both the primary and secondary controls must fail. The secondary or emergency controls may fail in the primary, secondary, or command modes. In this LAD, more reliable and effective controls can make the process safer.

The LAD for the EF3 (Exhibit 38). In this loss, the exposure exists because the process materials capable of a destructive reaction sufficient to release the fuel depend on control for containment. This destructive reaction is initiated as a result of the loss of process control designed to maintain that control through the operation of

Exhibit 38
Loss Through Destructive Reaction

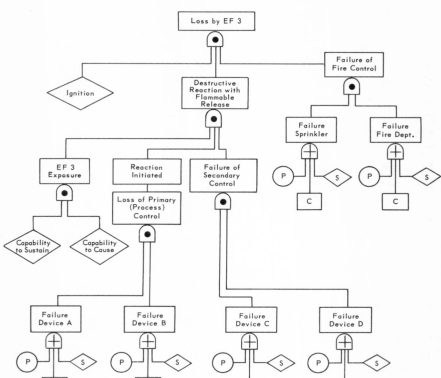

devices A and B. These devices may fail in primary, secondary, and command modes. The secondary or protective control devices, C and D, may be devices to stop the reaction rather than the relieving devices that are common to the EF2 loss mode. Ignition in this mode may be a separate event, or it may be concurrent with the reaction.

The three LADs shown in Exhibits 36, 37, and 38 cover those loss situations which result in fires and explosions. These losses are often dramatic, informing outsiders that perhaps the business is not as safe as it should be and that its presence threatens the community. Furthermore, these losses often claim lives. Both of these undesired effects provide the justification for pursuing the logic analysis (LAD) and identifying and correcting the causes of these losses.

Boiler and Machinery Losses

The other destructive mode of loss results from the misapplication of mechanical energy. The general LAD for losses arising out of the breakdown of boilers and machinery is shown in Exhibit 39. As before, there must be both the capability for the loss and the equipment failure before the loss can be experienced. The failure of the equipment can arise from a primary failure, such as fatigue or general deterioration, from a secondary failure from an event outside the design conditions of the equipment, or from a command failure. The command failure requires a failure of primary control devices and secondary protective devices. A typical command

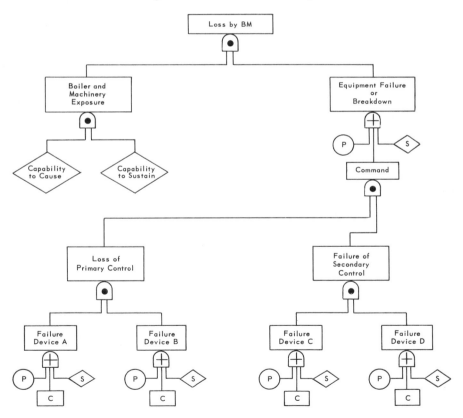

Exhibit 39
Loss Through Boiler and Machinery Breakdown

failure is overspeeding of a compressor-driver unit in which the governor (primary control) fails to control the speed and the overspeed trip (secondary protection) fails to trip the cutoff switch or control valve, permitting the machine to run to destruction. Again, each primary control and secondary protection device can fail in primary, secondary, or command modes.

Similar diagrams are possible for the nondestructive modes of business interruption (BI), material losses (ML), and personnel injuries (PI). These losses are usually experienced as part of the EF or BM modes, but sometimes a very costly loss can be sustained without any destructive result.

Nondestructive Losses

In the BI loss mode, the capability to *cause* the loss is readily available since any event that can cause the business flow to be interrupted has such capability. However, the capability to *sustain* the loss is not always available. For instance, in a business in which the product is boxed on a power-operated packaging line, a power failure will interrupt the flow of boxed product. If this occurs during a period of light demand, as in an off season, the interruption may not be serious enough to sustain a significant loss. But if the product is in great demand and requires full-time operation of the equipment, then any interruption of any significant time period will result in a significant loss. Certain equipment or process designs allow for down or nonoperating time, usually expressed as "percent onstream time" or operating time per calendar year. Periods of nonoperation within this design parameter cannot be classified as valid business interruption losses. However, since these design parameters allowing nonoperating time are usually arbitrarily defined, they should never be the excuse for ignoring the valid loss capability.

A typical example of a BI loss results from the interruption of the flow of essential raw material: The supplier of the commodity has a breakdown in his plant; the transportation system fails to deliver when expected; or a power failure occurs at the supplier's facility. These are secondary failures arising out of external events.

An example of material loses (ML) is the production of off-specification material; this is usually called a quality control problem. This produced material has to be marketed at a lower price or reprocessed. In either case, a loss of process control takes place.

Most personal injuries are not associated with any destructive loss of property. Although these could be analyzed by systems methods, such complicated analysis would not be justified for most injuries, especially those at lower severity levels. However, in a review of the process design, the potential injuries that might deserve analysis in detail would be (1) potential inhalation of toxic gases that are normally present in the process but might be released, without reaching a destructive loss, through loss of process control or a command failure, as during sampling; (2) potential thermal burns as a result of uninsulated hot pipes or equipment, or as a result of locating equipment in such a way that servicing it puts the worker in a hazardous position; or (3) potential strain injuries that can arise out of material handling or maintenance situations, for example when an employee is "instructed" to lift where a hoist could be provided, to move material where a conveyor or buggy could be provided, to strain where the position of the worker could be improved by better equipment location.

These three particular situations have been chosen simply because the injuries resulting from inhalation, burns, and strains are usually severe and potentially painful, serious, and costly. The experiences of the safety engineer and the maintenance supervisor are assets when reviewing the design of facilities, because they live with these problems daily.

This systems approach to loss situations, unlike the intuitive approach, demands that the capability to cause and the capability to sustain a loss be recognized as essential inputs. The rashness of such statements as "It can't happen here," "The vapor space of the tank is always too rich to burn," "The fire protection must be provided even though the maximum loss possible is far less than the cost of the protection" is exposed by the objective analysis of LAD.

The LADs illustrated are general diagrams typical for the particular loss modes. For a specific exposure, the diagram might be quite different even though the basic concepts would be retained. Diagrams of actual occurrences would reflect precisely the systems under consideration and specifically identify the controls. Detailed diagrams illustrate the need for controls to attain an effective design and expose the redundancy within control systems that makes the design inefficient and unnecessarily costly.

Quantifying Probability

Murphy's Law, often quoted, holds that if an event is possible and time is infinite, the probability of occurrence is a certainty. But sometimes Murphy's Law is applied without the condition of infinite time, permitting the inference that the occurrence of an event is highly probable. Occasionally this becomes the basis for extensive loss prevention activity beyond any justifiable position, and provides an excuse for doing nothing about a loss exposure that deserves attention. This is an illogical or unreasonable approach. Loss control decisions should be based on quantified probabilities and, therefore, require a practical method of quantifying probability.

Probability is the mathematical expression of the likelihood that an event will occur. This mathematical expression is a fraction falling between 0 and 1. It can be written as $0 \leq P \leq 1$ where 0 is absolute impossibility and 1 is absolute certainty.

Probability may be determined a priori or by observation. For example, given a perfectly balanced coin, we deduce a priori that when the coin is flipped, the probability of heads equals the probability of tails equals 0.5. A probability based on observation is determined by a number of trial flippings. In like manner, the probability of a baseball player's producing a hit at his next turn at bat is based on his batting average, which gives his past performance. Naturally, probability applies for the long run over many trials, but it may not hold for a specific trial.

Laws of Statistics

Statistics is the science of probabilities. There are laws, corollaries, conclusions—all of which may be found in the many textbooks on the subject.[3] Only those laws of statistics pertinent to the loss process and to the LAD will be expounded here.

[3] M. J. Moroney's *Facts from Figures* (Penguin Books, Inc., Baltimore, Maryland, 1965) is a very good text on statistics.

The *and* gate is used when all the input events must occur to bring about the output event (Exhibit 40). The probability of the output event is, therefore, the product of the probabilities of the input events. This explains the multiplication dot inside the gate symbol.

The *or* gate is used when only one of the input events is needed to produce the output event (Exhibit 41). The probability of the output event is the sum of the probabilities of the various input events. This explains the plus sign inside the gate symbol. Obviously, when more or less widely spread exponents are added, the sum will rarely exceed the largest of the inputs by one order of magnitude. Therefore, the largest input is used. In Exhibit 41, the sum of the relative probabilities is 0.1111, not significantly larger than 0.1, so the output relative probability is accepted as 0.1 or 10^{-1}. Of course if the sum indicates the next order of magnitude, then that order should be used.

This use of the sum of input probabilities to determine the probability of the output event is not mathematically rigorous; but because of the low probability values involved, the sum never reaches the value of one and therefore can be used effectively in the LAD.

As probabilities are assigned to input events, either through trial

Exhibit 40
Operation of the AND Gate

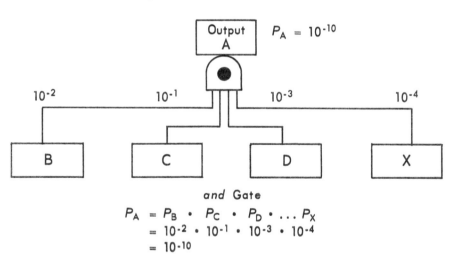

and Gate
$$P_A = P_B \cdot P_C \cdot P_D \cdot \ldots P_X$$
$$= 10^{-2} \cdot 10^{-1} \cdot 10^{-3} \cdot 10^{-4}$$
$$= 10^{-10}$$

Exhibit 41
Operation of the OR Gate

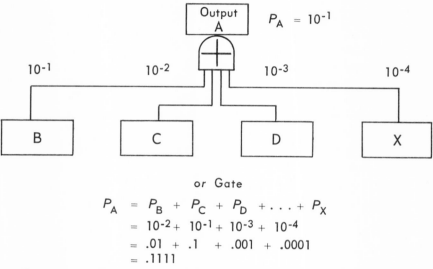

or Gate

$$P_A = P_B + P_C + P_D + \ldots + P_X$$
$$= 10^{-2} + 10^{-1} + 10^{-3} + 10^{-4}$$
$$= .01 + .1 + .001 + .0001$$
$$= .1111$$

or observation, the LAD can be solved mathematically for the output probability. Quantifying the LAD in this manner yields the probability of loss. Having both the loss potential and the loss probability, we can examine the balance. If this balance is unfavorable, we can make justifiable corrections.

Obviously and unfortunately, there is a limited supply of probability information for the various devices or inputs in the LAD. As in the case of personal injuries, accidents involving property have various severity classifications, and a probability relationship does exist between those various classifications. This probability relationship can be developed for a particular business by recording, over a long enough period of time to obtain a significant amount of data, the occurrence and the value of each loss. This was done in the case of a major chemical plant. The results are shown in Table 4.

This study, using ten years of incident history, shows the frequency ratio of the various levels of severity as well as the average time period between losses. The safety engineer can appreciate the similarity of groupings because these are used in the personal safety field. The hours between incidents is similar to man-hours, again a common term in the field of personal safety. Another way of stating

Table 4
PROPERTY LOSS SEVERITY PROBABILITIES

Financial Loss	Observed Frequency	Hours Between Losses	Severity
Over $1 million	0.1	10^5	Catastrophic
$100,000 to $1 million	1.0	10^4	Major
$25,000 to $100,000	10.0	10^3	Serious
$0 to $25,000	100.0	10^2	Minor

the time period is to say that a loss of catastrophic proportion (greater than $1 million) occurs every ten years or so, a major loss every year or so, a serious loss every month, and minor ones happen twice a week. The frequency of the so-called near-miss incidents can only be imagined.

Since incident frequency is the number of occurrences in a given time period, the relative probability of the incident occurrence can be expressed in the values of the denominator, the time period. For instance, if a major loss occurs once every 10,000 hours, the relative probability can be expressed as 10^{-4}, or we could say that the major loss has a relative probability rating of -4. Of course, any system of rating can be used if it is understood and useful to the individual or business. Using the system suggested, Table 5 shows the probability ratings for the various severity classifications.

To use these relative probability ratings in the LAD, the loss in-

Table 5
LOSS RELATIVE PROBABILITY RATINGS

Severity	Financial Loss	Frequency, Events/Hr.	Relative Probability Rating
Catastrophic	Over $1 million	10^{-5}	-5
Major	$100,000 to $1 million	10^{-4}	-4
Serious	$25,000 to $100,000	10^{-3}	-3
Minor	$500 to $25,000	10^{-2}	-2
Sub minor	$0 to $500	10^{-1}	-1

cident historical data for a particular business must be analyzed so that the proper relative probability can be assigned to the various types of loss events. It is important to understand that the relative probability rating was based on a study of the frequency of loss events regardless of classification, and that the rating was assigned to the various severity levels of actual losses.

The Tradeoff Level

The observed frequencies shown in Table 4 are an indication of the tradeoff value considered justifiable by management. Although management does not want any losses, a reasonable level is tolerated because it is felt that the expenditures or changes necessary for any significant improvement are not justified. The relative probability ratings shown in Table 5 indicate the accepted or tolerated tradeoff level. Any risk with a greater rating exceeds this tradeoff level and is termed hazardous. It is the responsibility of the safety engineer to analyze loss exposures for potential and for probability and to report with recommendations to management when these ratings are exceeded and the risk is hazardous.

This rating can be used for the various loss events in the LAD because it reflects the probability of the severity level, and a loss value can be ascribed to the various loss events through experience and judgment. These assigned ratings are shown in Table 6. For example, in the loss of process control (Exhibit 37), the loss requires the combined failure of several devices. If only device A should fail —and the frequency of such an occurrence might be quite high—the loss would be limited to the repair of the device, and the cost would be relatively low, giving a rating of -1 or at most -2. However, a coincidental failure of both A and B devices would have a higher loss result but a lower frequency, so the rating might be -3. Remember that each numerical change in the rating means a tenfold change in frequency.

As is the case with personal injury probability relationships, the numerical values may not be the same for all industries, plants, or businesses, but the relationship will exist. The capital value and the cost of interruption of the business and its subdivisions affect the numerical values of the losses in the various severity levels.

The Loss Control Program 153

The value of probabilities for loss analysis and risk evaluation can best be demonstrated through a practical example. Let us use a belt-driven motor-compressor unit with a receiver tank for surge before the air is delivered to various consumer units. The compres-

Table 6
PROBABILITY RATINGS FOR LOSS EVENTS

Description	Probability Rating
1. *Loss of control, process*	
(a) One failure: device, procedure, person	−2
(b) Two failures: device, procedure, person	−3
2. *Loss of control, mechanical or other*	
(a) and (b) as above.	
3. *Primary failure, containing equipment*	
(a) Massive release requiring only ignition	
(1) Heavy vibration or corrosion area	−3
(2) Light vibration or controlled corrosion	−4
(3) Nonvibrating—noncorrosive	−5
(b) Minor spill, ignitable only at source Reduce each item of (a) by one unit	
4. *Primary failure, mechanical equipment*	
(a) Major component of important equipment with extended disability (over one week) for replacement of available spare	−4
(b) Less critical breakdown with disability limited to 24 hours	−3
(c) Minor disability—installed spare existing or replacement of bearing, seals, etc.	−2
5. *Primary failure, electrical equipment*	
(a) Failure of distribution component with limited area interruption of several hours	−3
(b) Breakdown of major or critical drives	−4
(c) High voltage transformer disability	−4
6. *Loss of essential input, primary utilities to plant complex*	
(a) Extended disability over 24 hours	−5
(b) Limited disability of less than 24 hours	−4
(c) Flicker, maximum of a few minutes	−3
(d) Limited to one unit within plant	−2
7. *Primary ignition*	
(a) Adjacent or built-in source	−3
(b) Separated source	−4
8. *External event*	
(a) Freeze over extended periods	−4
(b) High winds greater than 125 mph.	−4
(c) Rains, floods, etc.	−4

Exhibit 42
Compressed Air System

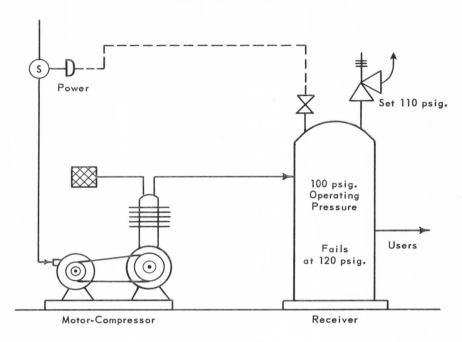

sor, diagrammed in Exhibit 42, is the nonlubricated type. The loss of the compressed air as a result of some interrupting event is a significant loss to the operation of the consumer units.

In this system, a pressure switch starts the motor-compressor at 90 psig and, after the pressure has been raised to 100 psig, the switch opens to stop the machine. A relief valve is set to open at 110 psig to protect the vessel from overpressure. The vessel, through some unconventional conditions (and for the sake of this demonstration) will fail destructively if subjected to pressures exceeding 120 psig.

By referring to the basic LAD (Exhibit 32), a loss in this system can be experienced in several modes. The methodical analysis of the loss exposure (the capability to cause and the capability to sustain a significant loss) will show that an EF loss would not be probable since there are no flammables available. A BM loss with accompanying business interruption is possible and should be analyzed further. The other modes are dependent on a BM loss, so they need not be considered. Although several modes of BM are possible—belt breakage and other component failures—the significant loss will be the rupture of the air receiver.

Exhibit 43 shows the LAD for the BM exposure. The capability to cause is the energy of the compressed air and the capability to sustain is the receiver and the surrounding equipment as well as the business interruption potential which will be realized if the receiver ruptures. For this example, a loss output of $100,000, including BI, has been estimated for which a relative probability rating of -3 is the barely acceptable tradeoff point.

A primary failure of the receiver requires that the pressure

Exhibit 43
The Boiler-Machinery Exposure

```
                    Compressor System      P = 10⁻⁵  (10⁻³ Acceptable)
                    Loss  $100M+
                          |
          ┌───────────────┴───────────────┐
                                        -5
    BM Exposure                    Destructive Failure
                                      of Receiver
          |                               |
     P = 1.0                       -7  ⟨P⟩─┴─⟨S⟩  -8
   ┌──────┴──────┐                        |
  C/S:         C/C:                    Command    -5
Receiver    Compressed                 P > 120
  etc.         Air                        |
                                ┌─────────┴─────────┐
                               -1                   -4
                        Loss of Primary Control   Failure Secondary Control
                               |                   |
                        Switch A Fails          RV Fails to Open at 110
                        to Open at 100
                               |                   |
                     -2 ⟨P⟩─┴─⟨S⟩ -6      -4 ⟨P⟩─┴─⟨S⟩ -7
                          [C]                    [C]
                           |                      |
                      ┌────┴────┐           ┌─────┴─────┐
                     -2        -1          -4          -7
                  Set Wrong  Valve Shut  Set Wrong  Obstructed
```

$P = 10^{-5}$ (10^{-3} Acceptable)

capability be less than that for which the receiver was designed—that is, there must be a failure in the design, defect in fabrication, deterioration in service, or similar fault. These failures have very low probabilities because pressure-vessel design and fabrication are well established with standards and codes, and compressed air service is not generally a corrosive situation. The relative probability rating of -7 is shown to indicate a very low probability; in fact, any number lower than -5 is an estimate with no incident record to confirm other than a very low probability. A secondary failure from some outside force that destroys the receiver is an even more remote possibility than a primary failure.

Destructive command failure will occur when the pressure in the receiver is made to exceed 120 psig. This requires the coincidental loss of both primary and secondary controls. The primary control is the pressure switch, which can fail to operate because of a primary or secondary or command failure. The switch may be commanded to fail by a wrong setting or by inadvertent closure of the valve in the pressure-sensing line. The relative probability rating of loss of the primary pressure control is equal to the highest value in this additive series (-1), or the inadvertent closure of the valve.

The secondary control or protective device is the relief valve, which may fail to open at 110 psig. This can happen as a result of a primary, secondary, or command failure. Here the command failure could be a wrong setting, which has a relative probability rating of -4, or an internal obstruction, which, in the case of a compressed air system, is a very remote possibility. The relative probability rating for the secondary or protective control is -4.

The relative probability rating for the vessel command failure is the product of the input ratings, and since these ratings are really the exponents of 10, we add the exponents to get -5. This is the governing probability for the destructive failure and, ultimately, for the $100,000 loss output. This design and protective system would be considered acceptable since the relative probability is much less than the tradeoff value of relative probability.

To better understand how these relative probabilities can be altered, the compressor system will be changed slightly. Exhibit 44 shows one such change—a block valve located on the inlet of the relief valve. Such a block valve is often installed in a relieving system so that the relief device can be serviced, tested, or replaced without shutting the system down.

In Exhibit 45, the LAD for this altered system is shown. The only change to the LAD is the addition of the third method of command failure in the secondary control branch, that of inadvertently closing the valve under the relief valve. The relative probability rating for inadvertently closing a valve is -1, which increases the probability of the receiver command failure to -2. This ultimately becomes the relative probability rating for the $100,000 loss output. This is not satisfactory because the relative probability of this system is greater than the acceptable tradeoff value. The best way to correct this exposure is to remove the valve, but when the value is essential, the probabilities can be made acceptable (in this example) by locking the valve in the open position. This reduces the probability of the inadvertent closure of the valve by at least one order of magnitude.

Exhibit 44
Insertion of a Block Valve

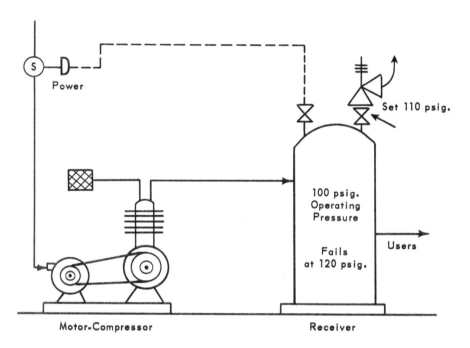

Exhibit 45
The Altered System

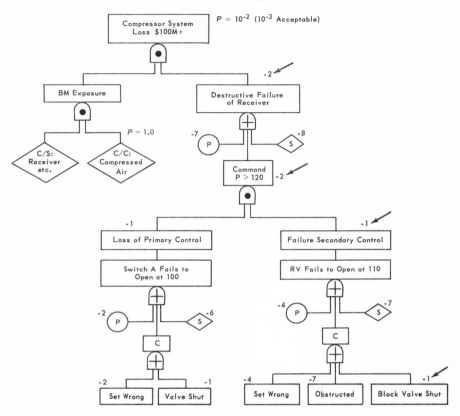

Another way to reduce the probabilities of the output loss would be to provide a redundant pressure switch, as shown in Exhibit 46. It should be noted that the two switch systems are entirely independent of each other. The LAD, Exhibit 47, shows the two independent switch systems under the primary control. The redundant switch reduces the probability of the primary control failure, and this is reflected in the reduced, now barely satisfactory relative probability of the loss output.

One rather common design fault in such systems is the failure to keep redundant controls fully independent. In Exhibit 48, both sensing lines to the pressure switches are connected to the same vessel nozzle through the same block valve. Closing one valve defeats both switch systems, and, as shown on the LAD (Exhibit 49), the relative probabilities are no longer acceptable.

This accident investigation before the fact is simply engineered property loss control.

Loss Control and Profitable Operations

A common excuse for avoiding the use of probabilities in evaluating probable maximum loss or attempting systems analysis is that probabilities for the variety of equipment, processes, and business failures are not available. But the approach in this book to probabilities does not purport to be rigorous, completely accurate, or infallible. And the particular numbers are not descriptive of any particular industry or business. It is presented as a reasonable and logical method by which one can get a handle on the problem. It

Exhibit 46
The Redundant Pressure Switch

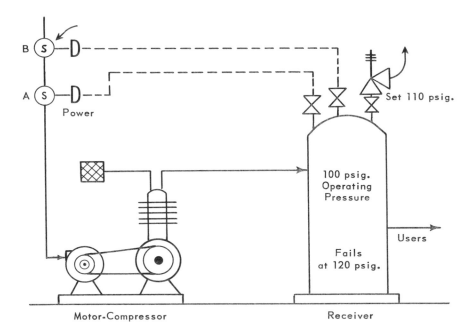

Exhibit 47
Two Switch Systems Under Primary Control

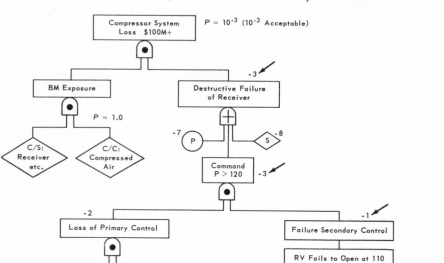

uses the history of reported losses and the experience and judgment of those concerned with the prevention of losses to assess the potentials and probabilities of losses in their business. This method is recommended as a means of calling management's attention to existing problems so that at least some of these problems will be corrected before an actual incident proves that we cannot afford the luxury of waiting for the probabilities and potentials to be established.

Loss control must be balanced with profitability. What would be gained if by making every operation foolproof so that no accident could happen, we were no longer capable of remaining in business? The point is that we must selectively attack loss exposures so that correction improves the safety and, ultimately, the profitability of the operation. This balance of profitability and loss control is simply the business decision of how many dollars should be expended toward the reduction of loss probabilities and potentials. The task is complex, but we know it is good business to preferentially apply the effort and dollars to the $100,000 loss that occurs annually, rather than to the $1 million loss that occurs once every 10 to 20 years. It follows that more effort should be expended on the various daily losses of $1,000 than on an annual $100,000 loss. The effort that reduces the frequency of the small losses will bring experience and awareness to bear on the larger loss.

It is within the responsibility of the loss prevention engineer to determine the loss potential and the loss probability. He can participate in the decision of profit expenditure by setting an objective priority on the probable loss of significant potential. But the management of the business must make the decision. How much easier

Exhibit 48
Two Sensing Lines on the Same Vessel Nozzle

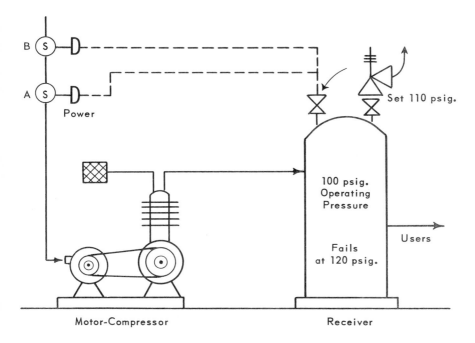

that decision will be when the manager knows the loss exposure, the probable maximum loss, and the probability of its occurrence.

The development of LADs for specific loss situations provides other advantages. The LAD enables us to visualize the loss process, and often we can see otherwise unrecognized causes or modes in which the loss might develop. The LAD points out the unfavorable probability and shows how to correct this unfavorable situation most effectively. When used after a loss has occurred, the LAD often locates unsuspected causes and indicates system corrections. The LAD highlights the need for adequate controls to maintain a favorable probability, but it also highlights the superfluity of control devices that do not significantly reduce the probability of loss.

Exhibit 49
Unacceptable Relative Probabilities

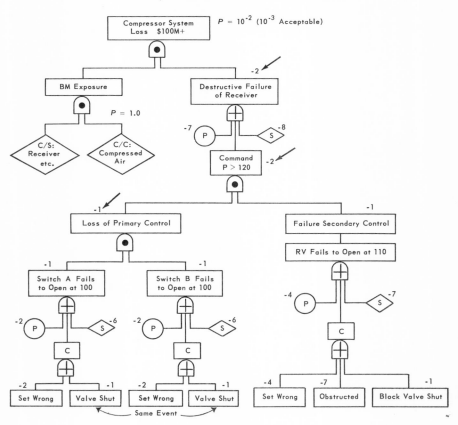

This is especially true of the secondary protective devices which are hung onto process equipment for an exposure that has a nonexistent or low probability while another exposure of high probability is ignored. The LAD also leads the designer to develop primary control devices of accurate response and insured reliability at points of critical exposure to make a process safer for those who operate that process.

This method of risk analysis is an opportunity for the safety engineer to broaden his contribution to the safety of the people and the facilities in his business. It enables the safety engineer to attain the technical competence demanded of staff specialists and to become the protector desired by the employees. This is no paradox, because the simple fact is that loss prevention is in the interest of both the welfare of the employees and the profitability of the business.

THE AUTHOR'S POSTSCRIPT

THIS book was written because I am convinced that loss prevention—a vital function of the business of producing goods and services—has been limited by a lack of measurement and control tools. Loss prevention will not gain the attention it deserves and it will not be used to full advantage until it develops discipline and scientific method. I do not expect that everything written in this book will find acceptance, nor do I imagine that a conclusive statement has been made. Rather, it is my humble hope that the tremendous progress made by industrial and business managements through the guidance of dedicated safety engineers in the interest of the safety, health, and security of the American working people shall be furthered by this presentation.

BIBLIOGRAPHY

Adams, Edward E., "The Nature of Loss Prevention," *ASSE Journal*, August 1966, Vol. XI, No. 8, pp. 17–20.

Allison, William W., "High Potential Accident Analysis," *ASSE Journal*, July 1965, Vol. X, No. 7, pp. 9–12.

———, "High Potential Accident Analysis: A Reappraisal," *Safety Maintenance*, December 1966, Vol. 132, No. 6, pp. 27–28.

"American Recommended Practice for Compiling Industrial Accident Causes," ASA Standard Z16.2-1941, Parts 1 and 2, American Standards Association, New York, New York.

"American Recommended Practice for Compiling Industrial Accident Causes," ASA Standard Z16.2-1962, Revision of Z16.2-1941, Parts 1 and 2, American Standards Association, New York, New York.

American Society of Safety Engineers, *Scope and Functions of the Professional Safety Position*, 850 Busse Highway, Park Ridge, Illinois.

Anderson, Kerme, "Pride: Do-it-Yourself Employee Motivation Program," *Public Relations Journal*, July 1964.

Andress, F. J., "The Learning Curve as a Production Tool," *Harvard Business Review*, Jan.–Feb. 1954, Vol. 32, No. 1, pp. 87–97.

Attaway, C. D., "Safety Sampling Program," Longhorn Division, Thiokol Chemical Corporation, DA-11-173-AMC-200(A), February 12, 1963.

Bird, Jr., Frank E., and George L. Germain, *Damage Control* (New York: American Management Association, Inc., 1966), pp. 21 and 45.

Blackman, A. C., "Professional Safety Engineering—A Need and Opportunity," *ASSE Journal*, May 1962, Vol. VII, No. 5, pp. 25–26.

Bowen, Gordon L., and Donald C. Whytock, *Economics of Safety* (Waterford, Conn.: Prentice-Hall, Inc., 1963), pp. 1–169.

Bower, G. W., "Don't Be Deceived by Frequency Rate," *Occupational Hazards*, September 1964, pp. 44, 46.

Brenner, Robert, and J. H. Mathewson, "The Principle of Accident Effect Reporting," *ASSE Journal*, January 1963, Vol. VIII, No. 1, pp. 9–14.

"British Engineers Ask: How Far Can the Accident Rate Be Cut?" reprint appearing in *ASSE Journal*, October 1962, Vol. VII, No. 10, pp. 27–30.

Brooks, Bert W., M.D., "An M.D.'s Use of Electronic Data Processing in Health and Safety," paper presented at 49th National Safety Congress and Exposition, October 16, 1961.

Browning, R. L., "Analyzing Industrial Risks," *Chemical Engineering*, October 20, 1969.

———, "Calculating Loss Exposures," *Chemical Engineering*, November 17, 1969.

———, "Estimating Loss Probabilities," *Chemical Engineering*, December 15, 1969.

———, "Finding the Critical Path to Loss," *Chemical Engineering*, January 26, 1970.

Burns, Robert K., "Management—A Systems Approach," *ASSE Journal*, December 1964, Vol. IX, No. 12, pp. 13–16.

Case, Harry W., "Safety Engineering and the Engineering College," *ASSE Journal*, March 1965, Vol. X, No. 3, pp. 17–19.

Cherry, Benjamin T., "The 'SPATS' Path to Construction Safety," *ASSE Journal*, September 1965, Vol. X, No. 9, pp. 18–21.

Conway, R. W., and A. Schultz, "The Manufacturing Progress Function," *Journal of Industrial Engineering*, Jan.–Feb. 1959, Vol. X, No. 1, pp. 39–54.

DeReamer, Russell, *Modern Safety Practices* (New York: John Wiley & Sons, Inc., 1958), pp. 293–312.

Dunlap, Jack W., *Manual for the Application of Statistical Techniques for Use in Accident Control* (Stamford, Conn.: Dunlap and Associates, Inc., June 1958), pp. 1–67.

Eng, G. H. C., "Safety Assessment—A Method for Determining the Performance of Alarm and Shutdown Systems for Chemical Plants," *Measurement and Control*, April 1968, Vol. 1, pp. 72–79.

Faggioli, Richard E., "Safety's Ups and Downs," *National Safety News*, August 1960, pp. 24–116.

Ferguson, Earl J., and James M. Daschbach, "SAF-HANS—A Positive Means to Improve Safety," *ASSE Journal*, February 1967, Vol. XII, No. 2, pp. 14–17.

Fitzgerald, E. B., "Quality—Whose Responsibility?" *Industrial Quality Control*, August 1965.

Griffin, Gerard O., "Little-known Facts About Injury Occurrence," *National Safety News*, May 1960, pp. 63, 138–141.

Grimaldi, John V., "Another Look at Stimulating Safety Effectiveness," *ASSE Journal*, April 1962, Vol. VII, No. 4, pp. 20–23.

———, "Appraising Safety Effectiveness," *ASSE Journal*, November 1960, Vol. V, No. 4, pp. 57–62.

———, "Management and Industrial Safety Achievement," *ASSE Journal*, November 1965, Vol. X, No. 11, pp. 9–14.

Halpin, James F., "Zero Defects in Retrospect," *Journal of American Society for Quality Control*, June 1966, Vol. 22, No. 12, pp. 669–670, 680–681.

Harvey, N. C., "Do Plant Safety Committees Serve a Useful Purpose? British Safety Officer Says No!" *ASSE Journal*, May 1964, Vol. IX, No. 5, pp. 17–20.

Heinrich, H. W., *Industrial Accident Prevention* (New York: McGraw-Hill Book Company, third edition, 1953), p. 24.

Hirschmann, W. B., "The Learning Curve," *Chemical Engineering*, March 30, 1964, pp. 95–100.

———, "Profit from the Learning Curve," *Harvard Business Review*, Jan.–Feb. 1964, p. 125.

Howell, J. M., and Lee Johnson, "Statistical Control of Accidents," *The Iron Age*, July 18, 1946, Vol. 158, No. 3, pp. 56–59.

Kahn, Louis B., "A Statistical Model for Evaluating the Reliability of Safety Systems for Plants Manufacturing Hazardous Products," *Technometrics*, August 1959, Vol. 1, No. 3, pp. 293–307.

Klingel, A. R., and O. C. Haier, "SOHIO Serious Injury Index," *National Safety News*, November 1956.

Kolodner, Herbert J., "The Quantification of Safety," *ASSE Journal*, March 1965, Vol. X, No. 3, pp. 9–12.

Levens, Ernest, "Systems Analysis—A Powerful Tool for Accident Prevention," *National Safety Congress Transactions, Chemical and Fertilizer Industries*, 1967, Vol. 5, pp. 25–32.

Logan, P. W., "Engineering—Tomorrow's Basic Component for Accident Prevention," *ASSE Journal*, November 1966, Vol. XI, No. 11, pp. 7–11.

Martin, J. A., and G. B. Wheeler, M.D., "Safe-t-Scores: A New Measurement for Safety," *Occupational Hazards*, April 1962, Vol. 24, No. 4, pp. 35–39.

Mazel, J., "Setting Up a Zero Defects Program," *Factory*, July 1965, McGraw-Hill Book Company.

McFarland, Harold S., "The Concept of Accident Prevention as a Basic

Management Function," *ASSE Journal*, December 1963, Vol. VIII, No. 12, pp. 11–13.

"Measuring Industrial Safety Performance" (a symposium report), *National Safety News*, December 1966, Vol. 94, No. 6, pp. 36–41.

Meyer, Joseph J., "Statistical Sampling and Control for Safety," *Industrial Quality Control*, June 1963.

Moroney, M. J., *Facts from Figures* (Baltimore, Maryland: Penguin Books, Inc., 1965).

Mueller, Robert Kirk, *Effective Management Through Probability Controls* (New York: Funk & Wagnalls Company, 1958), pp. 174–177.

National Safety Council, *Accident Facts;* booklet published annually.

———, *Work Injury Rates*, 1967 Edition, p. 26.

Pape, W. C., U.S. Dept. of Int., and T. J. Creswell, Federal Aviation Agency, "Safety Aids Decision Making," unpublished paper.

Papers on fault tree analysis from the proceedings of the System Safety Symposium, Seattle, Washington, June 8–9, 1965, cosponsored by The Boeing Company and The University of Washington:

a. Mearns, A. B., "Fault Tree Analysis: The Study of Unlikely Events in Complex Systems," Bell Telephone Laboratories, Whippany, N.J.
b. Feutz, R. J., and T. A. Waldeck, "The Application of Fault Tree Analysis to Dynamic Systems," The Boeing Co., Seattle, Washington.
c. Haasl, D. F., "Advanced Concepts in Fault Tree Analysis," The Boeing Co., Seattle, Washington.

Payne, Charles L., "APEX—A System for Rating Accident Prevention Effort," *National Safety News*, September 1966, Vol. 94, No. 3, pp. 38–40.

Peters, G. A., and F. S. Hall, "Human Error: Analysis and Control," *ASSE Journal*, January 1966, Vol. XI, No. 1, pp. 9–15.

———, "Design for Safety," *Product Engineering*, September 13, 1965.

———, "System Safety Engineering as a Technical Discipline," paper presented at the Ergonomics and Aerospace Joint Session of the American Industrial Hygiene Conference, April 26–30, 1964, in Philadelphia, Pa.

Pollina, Vincent, "Safety Sampling," *ASSE Journal*, August 1962, Vol. VII, No. 8, pp. 19–22.

Preddy, D. L., "Guidelines for Safety and Loss Prevention," *Chemical Engineering*, April 21, 1969, pp. 94–108.

Recht, J. L., "Systems Safety Analysis: The Fault Tree," *National Safety News*, April 1966, Vol. 93, No. 4, pp. 37–40.

———, "Systems Safety Analysis: Failure Mode and Effect," *National Safety News*, February 1966, Vol. 93, No. 2, pp. 24–26.

———, "Systems Safety Analysis: An Introduction," *National Safety News*, December 1965, Vol. 92, No. 12, pp. 37, 38, 122.

Robert, Marcel, "Appreciation from Geneva—'Industry's Voluntary Safety Programs Are Key to Progress in Accident Prevention,' Says Safety Chief for ILO," *Engineering for Safety*, October 1959, pp. 8–10.

Rockwell, Thomas H., "A Systems Approach to Maximizing Safety Effectiveness," *ASSE Journal*, December 1961, Vol. VI, No. 6, pp. 17–22.

———, "Safety Performance Measurement," *Journal of Industrial Engineering*, January–February 1959, Vol. 10, pp. 12–16.

Rockwell, Thomas H., and Larry R. Burner, "Information Seeking in Risk Acceptance," *ASSE Journal*, February 1968, Vol. XIII, No. 2, pp. 6–10.

Ruddick, Clyde C., "The Value of Good Records and How to Use Them," paper presented at Accident Prevention Fundamentals Session, National Safety Congress, Chicago, October 23, 1957.

"Safety Committees—Pro and Con," *ASSE Journal*, July 1964, Vol. IX, No. 7, pp. 6–8.

Safety and Loss Prevention Guide: "Hazard Classification and Protection," Second Edition, The Dow Chemical Company, March 16, 1966.

Satterwhite, Henry G., and Robert M. LaForge, "A Comparison of Three Measures of Safety Performance," *ASSE Journal*, March 1966, Vol. XI, No. 3, pp. 9–15.

Schowalter, E. J., M.D., "A Year's Trial with a New Safety Measurement Plan," a paper presented at the 9th Annual Western Industrial Health Conference, Oct. 9, 1965.

Shiskin, Julius, "Electronic Computers and Business Indicators," Occasional Paper 57, National Bureau of Economic Research, Inc., 1957.

Smith, H. M., *Guide to Fire Prevention in the Chemical Industry*, British Chemical Industry Safety Council of the Chemical Industries Association, Ltd. (W. Heffer & Sons, Ltd., Cambridge, England).

Stresau, Ann, "Accident Rates and Changes in the Employment Level," *National Safety News*, August 1961, Vol. 6, No. 3, pp. 34–38.

———, "Safety—Reliability," *ASSE Journal*, January 1966, Vol. XI, No. 1, p. 16.

Tarrants, William E., "The Application of Inferential Statistics to the Appraisal of Safety Performance," Office of Industrial Hazards, U.S. Department of Labor, March 31, 1966, pp. 1–18.

———, "Applying Measurement Concepts to the Appraisal of Safety Performance," *ASSE Journal*, May 1965, Vol. X, No. 5, pp. 15–22.

———, "New Approaches to Industrial Safety Performance Measurement," Office of Industrial Hazards, U.S. Department of Labor, October 27, 1966.

Terry, Jr., S. B., *Fundamentals of Boiler and Machinery Insurance*, pamphlet published by The Hartford Steam Boiler Inspection and Insurance Company, Hartford, Conn.

Turner, A. W., "The ASA Disabling Frequency Rate Is *Not* a Good Measure of Safety Performance," *ASSE Journal*, January 1963, Vol. VIII, No. 1, p. 4.

Tuz, P. J., and N. J. DeGrazia, "The Socio-Economic Implications of Industrial Safety," *ASSE Journal*, September 1967, Vol. XII, No. 9, pp. 21–24.

United States of America Standards Institute (formerly American Standards Association), Z16.1, "Method of Recording and Measuring Work Injury Experience," 1967, pp. 7–28.

Voland, L. J., "Beyond Frequency and Severity," *National Safety News*, December 1962, Vol. 86, No. 6, pp. 28, 68.

Wallace, Earl R., "The ASA Disabling Injury Frequency Rate Is a Good Measure of Safety Performance," *ASSE Journal*, November 1962, Vol. VII, No. 11, pp. 14–16.

Wright, Lawrence T., "Try This Analog for Solving Your Fire-Water Problems," *Hydrocarbon Processing & Petroleum Refiner*, September 1962, Vol. 41, No. 9, pp. 274–278.

Wright, T. P., "Factors Affecting the Cost of Airplanes," *Journal of the Aeronautical Sciences*, February 1936, Vol. 3, No. 4, pp. 122–128.

"The Z16.1 Code Does Not Deceive . . ." (prepared by an anonymous member of the ASSE editorial board), *ASSE Journal*, November 1966, Vol. XI, No. 11, pp. 12–13.

INDEX

Accident prevention: injury effect measurement, 80–113; injury report, 48–79, 120; loss process, 33–38, 133, 134; management, 18–19; manager and, 9–10; performance measurement techniques, 20–32, 114–117; performance measuring system, 39–47; and safety engineer, 12–19; *see also* Loss control
Accident probability, 69–72
Accidents, *see* Injuries
Allis-Chalmers International, 83
Allison, William W., 29, 69
Aluminum Company of America, 29
American Petroleum Institute, 130
American Society of Safety Engineers, 14, 15, 23
American Standards Association, 20; Z16 Standard, 23
"Another Look at Stimulating Safety Effectiveness" (Grimaldi), 23, 84
"Applying Measurement Concepts to the Appraisal of Safety Performance" (Tarrants), 22, 30
"Appraising Safety Effectiveness" (Grimaldi), 84

Bell Telephone Laboratories, 31
Bird, Frank E., Jr., 43
Blackman, A. C., 23
Boeing Company, 31
Brenner, Robert, 25
Business interruption loss, 120–128
Business Process Safety Review, 130–132

Chrysler Corporation, 28
Control charts, 16, 79, 87, 96, 108, 126, 128; injury report, 57–68, 72–74; managers, 64–66, 67; rolling, 66–68, 72; safety engineers, 65, 78, 113; seasonal fluctuation, 61–65; standard deviation, 58–61; supervisors, 64–67; as trend charts, 72–74
Cost control, 16
Creswell, T. J., 19
Critical Incident Technique, 29–30, 114
Cumulative sums techniques, 84–85, 96, 113, 126

Damage Control (Bird and Germain), 43

171

172 INDEX

DeGrazia, N. J., 83
DeReamer, Russell, 22, 69–72
Disabling Injury Index, 23–24
Du Pont Safety Control Program, 28

Effective Management Through Probability Controls (Mueller), 21
Eng, G. H. C., 133

Facts from Figures (Moroney), 148
Failure Mode and Effect Method, 31–32
Fault Tree Analysis, 31

Germain, George L., 43
Griffin, Gerard O., 83
Grimaldi, John V., 22, 23, 83, 84
Group insurance, 24
Guide to Fire Prevention in the Chemical Industry, 131

Haier, O. C., 26, 70
Hall, Frank S., 32, 133
"Hazard Classification and Protection," 131
Hazard Survey of the Chemical and Allied Industries, 131
Heinrich, H. W., 42
High Potential Accident Analysis, 29
"High Potential Accident Analysis" (Allison), 29, 69
Hirschmann, Winifred B., 91, 93

Industrial Accident Prevention (Heinrich), 42
Injuries: accident probability, 69–72; disabling, 13, 21–24; and measurement techniques, 20–32; seasonal, 61–63, 81; serious injury frequency (SIF), 71–72; total injury frequency (TIF), 71–72; types of, 63, 81, 85–87
Injury effect measurement, 80–113; cumulative sums techniques, 84–85; historical, 81; learning curve, 90–113; Multiple Linear Regression (MLR) techniques, 96–113; rolling average method, 87–90, 113; safety engineers, 113; seasonal, 81; Seasonally Adjusted Economic Series, 90; Weighted Severity Index, 85–87
Injury frequency, 126
Injury report, 48–79, 120; accident probability, 69–72; control chart, 57–68, 72–74; employee attitudes, 48–50; forms, 51–57; manager and, 76–79; performance chart, 65–66; safety engineers, 51; serious injury frequency (SIF), 74–76; supervisors, 49–51, 54, 78; total injury frequency (TIF), 74–76; value on data, 49–51
Inspectors, 115, 117
Insurance: group, 24; workmen's compensation, 12, 24

Kahn, Louis B., 133
Klingel, A. R., 26, 70
"Knowing's Not Enough" (film), 65

Learning curve, 90–113
Logic analysis, 16
Loss: business interruption, 120–128; destructive modes of, 137–138, 145–146; exposure, 33–34, 133, 134; incident, 34–35, 40–43; nondestructive modes of, 138–139, 146–148; personal injury, 13, 21–24, 40, 128, 139, 147; potential, 114–119; process, 33–38, 133, 134; property, 41, 43, 120–128; types of, 40–43
Loss analysis diagram (LAD), 134–137, 140–144, 147–152, 155, 157, 158, 162, 163; statistics pertinent to, 148–152
Loss control: importance of, 10–11; injury effect measurement, 80–113; injury report, 48–79, 120; loss control program, 129–163; loss exposure, 33–34, 133, 134; loss incident, 34–35, 40–43; loss process, 33–38, 133, 134; management, 18–19; and manager, 9–10; performance measurement techniques, 20–32, 114–117; performance measuring system, 39–47; property and business interruption, 120–128; and safety engineer, 12–19; *see also* Accident prevention

Loss control program, 129–163; Business Process Safety Review, 130–132; destructive loss, 137–138; 145–146; laws of statistics, 148–152; loss analysis diagram (LAD), 133–137, 140–144, 147–152, 154, 155, 157, 158, 162, 163; nondestructive loss, 138–139, 146–148; profitable operations, 159–163; quantifying probability, 148; systems safety analysis, 132–134; tradeoff level, 152–159

Loss exposure, 33–34, 133, 134

Loss incident, 34–35, 40–43

Loss process, 33–38, 133, 134

Loss report, 43–47; coverage of, 46–47; managers and, 45; problems of, 45–46; safety engineers, 46, 121–122

Management: accident prevention and, 18–19; control systems, 14, 39–47; cost control, 16; and safety engineer, 14

Managers, 9–11, 16, 17, 121, 128, 132, 133; control chart, 64–66, 67; and injury report, 76–79; and loss control, 9–10; loss report, 45; sampling, 115, 116

Manufacturer's Chemists Association, 130

Martin, James A., 24–25

Mathewson, J. H., 25

Minuteman Program, 31

Modern Safety Practices (DeReamer), 22, 69

Monsanto Company, 28, 42, 83, 90

Moroney, M. J., 66, 148

Mueller, R. K., 21

Multiple Linear Regression (MLR) techniques, 96–113

Murphy's Law, 148

National Fire Protection Association, 130

National Safety Council, 21, 54, 130

National Safety News, 70

Occupational Hazards, 24

Olin Mathieson Chemical Corporation, 27

Pape, W. C., 19

Performance measurement techniques, 30–32; Disabling Injury Index, 23–24; inspectors, 115, 117; Safe-t-Scores Method, 24–26; safety engineers, 18–19, 20, 23, 25, 27, 31–32; sampling, 27–30, 114–115; Serious Injury Index, 26–27; systems analysis, 31–32; Z16.1 Code, 20–23, 24, 78

Performance measuring system, 39–47; loss incident, 40–43; loss report, 43–47; probability relationship, 43

Personal injury loss, 13, 21–24, 40, 128, 139, 147

Peters, George A., 32, 133

Poisson distribution, 66

Potential loss, 114–119

Pratt and Whitney Connecticut Atomic Nuclear Engine Laboratory, 24

"Principle of Accident Effect Reporting, The" (Brenner and Mathewson), 25

Probable maximum loss (PML), 134, 135

"Professional Safety Engineering—A Need and Opportunity" (Blackman), 23

Property loss, 41, 43, 120–128; frequency, 126–127; incident form, 122–125

Rockwell, Thomas H., 22, 39

Rolling average method, 87–90, 113

Rolling control charts, 68–69, 72

"Safe-t-Scores: A New Measurement for Safety" (Martin and Wheeler), 24

Safe-t-Scores Method, 24–26

"Safety Aids Decision Making" (Creswell and Pape), 19

Safety engineers, 10, 12–19, 47, 90, 114, 119, 129, 133, 150, 163; control chart, 65, 78, 113; cumulative sums techniques, 113; injury effect measurement, 113; injury report, 51; investigations, 17–18; learning curve, 113; loss report, 46, 121–122; and management, 14; past performance, 12–14; performance

174 INDEX

Safety engineers (continued) measurement techniques, 18–19, 20, 23, 25, 27, 31–32; professionalism, 14; sampling, 115; space-age techniques, 15–16
Safety and Loss Prevention Guide, 131
Sampling, 27–30; managers, 115, 116; as performance measurement, 114–115; safety engineers, 115; supervisors, 116
Schowalter, E. J., 23
"Scope and Functions of the Professional Safety Position," 15
Seasonally Adjusted Economic Series, 90
Serious injury frequency (SIF), 71, 80–81, 88, 92, 95, 96, 97, 108–112 *passim;* injury report, 74–76
Serious Injury Index, 26–27, 87
Shiskin, Julius, 90
Sohio Serious Injury Index Method, 70
Standard Oil Company, 26
State Workmen's Compensation Act, 85
"Statistical Model for Evaluating the Reliability of Safety Systems for Plants Manufacturing Hazardous Products, A" (Kahn), 133
Stresau, Ann, 83
Supervisors, 18, 119, 128; control chart, 64–67; injury report, 49–51, 54, 78; loss report, 121; sampling, 116
Systems analysis, 31–32

"Systems Approach to Maximizing Safety Effectiveness, A" (Rockwell), 22, 40
Systems safety analysis, 132–134
"Systems Safety Engineering as a Technical Discipline" (Hall and Peters), 133

Tarrants, William E., 22, 29–30, 114
Thiokol Chemical Corporation, 28
Total injury frequency (TIF), 71, 80–81, 87, 88, 90, 97, 108–111; injury report, 74–75
Tuz, P. J., 83

United States of America Standards Institute, 130; Z16.2 Standard, 54
U.S. Bureau of Labor Statistics, 83

Weighted Severity Index, 85–87
Western Electric Company, Incorporated, 23
Wheeler, Gordon B., 24–25
Work orders, 118, 126–128
Workmen's compensation insurance, 12, 24
Wright, T. P., 91

"Year's Trial with a New Safety Measurement Plan, A" (Schowalter), 23

Z16 Standard, 23
Z16.1 Code, 20–23, 24, 78
Z16.2 Standard, 54
Zero Defects Program, 30

ABOUT THE AUTHOR

CHARLES L. GILMORE is the Superintendent of Loss Prevention at the Texas City, Texas, plant of the Monsanto Company. He is responsible for the administration and direction of the loss prevention program, which includes personnel safety, property loss control, fire protection, and environmental hygiene.

Mr. Gilmore graduated from Rensselaer Polytechnic Institute with a bachelor of chemical engineering degree. He was then employed by the Standard Oil Company of New Jersey as a production supervisor in the olefin cracking and synthetic butyl rubber operations at Baton Rouge, Louisiana. At the conclusion of World War II, he joined the staff of the Hagan Corporation in Pittsburgh, Pennsylvania, as a research engineer in boiler-water treatment and later became plant superintendent of the Ellwood City, Pennsylvania, plant. In 1951 he began his employment with the Monsanto Company as a production supervisor of the vinyl chloride monomer plant. In 1956, he assumed his present position.

Mr. Gilmore is a professional engineer registered in the states of Pennsylvania and Texas. He is a Member of the American Society of Safety Engineers, a member of the General Advisory Committee for the Texas Occupational Safety Board, and a charter member and vice-chairman of the Occupational Safety Committee of the Texas Chemical Council. He was also general chairman of the Texas City Industrial Mutual Aid System. He has been a speaker at many state and national safety conferences and has published several articles on loss prevention and disaster planning for leading periodicals in his field.